Katherine C. Reeves

D0206288

Other Books by Donald Gropman:

Say It Ain't So, Joe

Other Books by Donald Gropman:

Say It Ain't So, Joe

Comet Fever

Donald Gropman

with Kenneth Mirvis

A FIRESIDE BOOK

Published by Simon & Schuster, Inc.   New York

Copyright © 1985 by Donald Gropman
All rights reserved
including the right of reproduction
in whole or in part in any form
A Fireside Book
Published by Simon & Schuster, Inc.
Simon & Schuster Building
Rockefeller Center
1230 Avenue of the Americas
New York, New York 10020
FIRESIDE and colophon are registered trademarks of Simon & Schuster, Inc.
Designed by Karolina Harris
Manufactured in the United States of America

1 3 5 7 9 10 8 6 4 2

Library of Congress Cataloging in Publication Data
Gropman, Donald.
Comet fever.

"A Fireside book."
Includes index.
1. Halley's comet—History.   I. Mirvis, Kenneth.
II. Title.
QB723.H2G76   1985   523.6′4   85-12966
ISBN: 0-671-60307-8

Acknowledgments

In tracking down the tantalizing story of the sacrificial virgin and the Oklahoma comet cult, many people offered assistance. Among them were Eleanor Landon of the Oklahoma Historical Society; columnist Kent Ruth of the *Daily Oklahoman*; Connie Halton and Ted Jones of the Woods County Sheriff's Office; Deputy Mike Zarilla of the Alfalfa County Sheriff's Office; Sheriff Jerry Metz of Major County; Laurie Thorne of the Alva, Oklahoma, Public Library; Virgil Pruitt; Bradley Gernand and Don Mustain Jr. of the Western History Collections of the University of Oklahoma at Norman; historians Howard Dominick and Ken Miller; Fred Bingamon of the Oklahoma Department of Corrections; and Eunice S. Estle, editor and publisher of the *Woods County News*.

Physicist Richard A. Muller of the University of California at Berkeley and the Lawrence Berkeley Laboratory offered some background on his Nemesis Theory. Special thanks are due to Professor Emeritus Fred Whipple of Harvard University, former director of the Harvard Smithsonian Astrophysical Laboratory. Dr. Whipple spent several hours patiently answering rudimentary questions about the science of com-

· · · · · · · · · · · · · · · · · ·

ets. Both of these scientists offered substantial assistance, but they are in no way responsible for the contents of this book.

Without the assistance of Ruth S. Freitag of the Science and Technology Division of The Library of Congress, the illustrations in this book would have been less delightful. Ms. Freitag researched and photocopied many of the 1910 comet images that are reproduced here. Valuable help in the search for illustrational materials was also offered by Wallace Dailey, curator of the Theodore Roosevelt Collection at the Houghton Library, Harvard University, and Bistra Lankova, reading room supervisor of Harvard's Theatre Collection. Pamela Ross of the Harvard Law Library arranged for the speedy arrival of newspaper microfilm.

Thanks are also due to John Pierce of Yankee Publications for his early support of this project; David Dortort for general encouragement; Virginia Risse for clipping comet-related news items; Laura Eastment for Spanish translations; Michael Connolly for steadfast support as literary agent for this project; Sonya Gropman for scrupulous reading of newspaper microfilm; Adam Gropman for unique questions and comic relief; and my wife Gabrielle Rossmer for French and German translations, photo research, photography, and infinite patience with a preoccupied writer.

These acknowledgments would be incomplete without public recognition of two of the main characters in the cast. Editor Herb Schaffner of Fireside Books brought an unmatched level of enthusiasm, insight and hard work to this project; he made editing a pleasurable process. And finally, the most special thanks are offered to Cathy Hemming, vice-president and publisher of the Trade Paperback Group of Simon & Schuster. Cathy's unwavering belief in this project made it possible. Without her, there would have been no book.

D.G.

Dedication

In memory of
Stephen H. Rossmer
and
Max W. Gropman

Contents

• • • • • • • • • • • • • • • •

Part Four · Our Rendezvous with Halley

Introduction

For a few days in May of 1910 the world went crazy. Seized by panic and fear, millions of ordinary people performed extraordinary deeds.

In Constantinople, tens of thousands of terrified people packed services at mosques and churches. Hordes of Londoners roamed the streets all night, waiting for morning. Mexicans overflowed their churches and cathedrals, and huge throngs gathered on hillsides for prayer sessions. In New York, Paris, St. Petersburg, Berlin, Lausanne, and other metropolitan centers around the globe, the elite gathered for soirées on hotel rooftops, or ascended into the night sky in hot-air balloons.

In the American midwest, farm families huddled in cyclone cellars. In South Africa, cautious mining engineers took their families to the bottoms of the mines. But in the Appalachian coal fields the behavior was the opposite—fearful miners refused to descend into the mines. In many of the world's ports, dockworkers refused to work. Orchards, fields, and farms fell silent. City dwellers everywhere sealed their windows and chimneys with tape and rags, trying to make them as airtight as possible.

• • • • • • • • • • • • • • • • •

The sale of bottled oxygen soared, and there was an absolute run on binoculars and telescopes.

On a more tragic note, a wave of suicides swept around the world. A California man crucified himself. And out of Oklahoma came the story of a fanatical cult that tried to sacrifice a virgin.

This mass hysteria was caused by Halley's comet during its last visit in the spring of 1910. The world's response was not unprecedented. People have always reacted strongly to a visit by a comet, particularly to comet Halley, one of the two or three most legendary objects in the universe. But what was most unusual about the world's response was its intensity. The epidemic of comet fever that raged in May of 1910 was one of the worst in history.

Comets are mysterious and powerful objects. To this day, scientists do not know with certainty what exactly a comet is, nor do they know where comets come from. During Halley's 1985–86 visit, space-age technology will be employed in an attempt to solve some of the most basic questions and riddles about comets. But whatever their physical composition and origin, one thing is certain: comets have always stimulated the human imagination.

Until the dawning of the Age of Enlightenment in the eighteenth century, comets were universally regarded as omens of doom and disaster, even by the most sophisticated of people. When a comet appeared, whole populations cowered in superstitious terror, while scholars, clerics, and astrologers tried to provide supernatural explanations.

By the time we reached the twentieth century, most enlightened observers believed the age of superstition was dead, or at least in its last hours. Before Halley returned for its first visit in the twentieth century, the common opinion among the enlightened was that the populace of 1910 would be thrilled by the sight of the comet, not terrified as their ancestors had been in the dark past. As things turned out, this faith in enlightened rationalism was overly optimistic. Like the premature reports of Mark Twain's death, it was greatly exaggerated.

• • • • • • • • • • • • • • • • • • •

The belief in rationalism was not a wild flight of fancy, it was well grounded in the experience of the early twentieth century. Science and technology had progressed at a startling rate. In fact, the rate at which inventions and discoveries were being made was so rapid that some observers announced that humankind had reached the limits of scientific and technological progress. Ironically enough, it was two small pieces of this progress that led the world backward into a frenzied regression of mass hysteria.

When Halley came into telescopic view late in 1909, recent advances in spectrographic analysis allowed astronomers to analyze its chemical composition. They discovered that the comet's tail contained cyanogen gas, a lethal poison. Prior to this, cyanogen gas had been found in the tail of at least one other comet, so this new discovery was announced without fanfare.

A short time later another discovery was made. Using formulas based on the most advanced understanding of celestial mechanics, astronomers plotted Halley's orbit and learned that the Earth would pass through the comet's tail on the night of May 18–19. The announcement of this discovery, added to the earlier discovery of the cyanogen gas, was the direct cause of the panic that followed. All the reassuring statements issued by scientists and editors around the world could not prevent the great mass of the world's population from fearing that the poison gas would kill everyone on Earth.

There was probably no way the mass of people could have been relieved of their anxieties. News of the approaching encounter with the poison tail seems to have triggered an instinctive response, the kind of response that is not very susceptible to the power of reason. Some theorists have even argued that the entire human race carries a subconscious memory of actual comet disasters that occurred in the ancient past. These memories are stirred by the sight of a comet. The specific events of 1910 brought them back to life.

The comet photographed from Argentina in May 1910. (From The
Library of Congress.)

Part One

From Souls
to Snowballs

Cometa

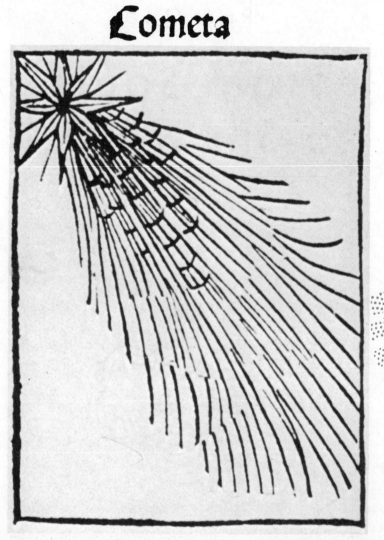

Halley's apparition of 684 A.D. as recorded in the Nuremberg Chronicle. A more imaginative description comes to us from the sixteenth century: "On both sides of the rays of this comet were seen a great number of axes, knives, and blood colored swords among which were a great number of hideous human faces with beards and bristling hair." (From The Library of Congress.)

Chapter 1

Cometology
Before Halley

S **Aristotle's Perfect, Unchanging Universe**

ince our earliest days, we have always turned our eyes upward to the heavens. The awesome mysteries of sun, moon, and stars have had the most profound effect on the collective imagination and intellect of humankind. Astronomy is one of the oldest sciences, and the ancient writings of such civilizations as Babylonia, India, China, and Egypt contain astronomical data and speculation. It is believed that eclipses were accurately predicted as long ago as 4000 years.

Until the time of the Greeks, however, much of the observational work we would now call "astronomy" was concerned with the practical matters of mapping the heavens, identifying the stars, and developing the instruments needed for these kinds of endeavors. To a great extent, the data collected about the heavenly bodies were used by astrologers to interpret how these bodies affected people and to predict who would fall under their usually baleful influence, and when.

As they were in most of the sciences, the Greeks were the first to introduce theoretical concerns to the study of the heavens, out of which, eventually, grew a model of the universe that consisted of a series of perfect concentric spheres with the Earth in the center.

• • • • • • • • • • • • • • • • •

Over 2000 years ago, Aristotle posited a theory that the universe was divided into two regions, one beneath the moon (the sublunar universe), and the other extending to include everything beyond the moon. In Aristotle's cosmology, the things beneath the moon are subject to change, and the things beyond, including our sun and the other stars, are eternally unchanged.

Aristotle's geocentric model was later incorporated into the cosmology of the Church, and remained a matter of dogma for centuries. It was a stifling concept, because the "truth" as described by dogma did not fit the actual observations made by new generations of scientists. Aristotle's theories, according to Bertrand Russell, held back the development of science for 2000 years.

Among other things, Aristotle's ideas inhibited the development and acceptance of the heliocentric model of the Solar System. Men who said the sun was the center of the Solar System were forced to recant (Galileo), or were burnt at the stake (Bruno). Among its other negative effects, Aristotle's theory inhibited the development of the scientific study of comets.

Since according to Aristotle, the spheres of the universe beyond the moon were perfect and immutable, where nothing ever happened and nothing ever changed, it followed that comets, being erratic, unpredictable, and apparently transitory, must exist between the moon and the Earth. Thus went the reasoning implicit in Aristotle's view.

Following the official position of the scientific/theological establishment, observers of comets devised interpretations to fit the rules. From ancient times through the Middle Ages, comets were variously held to be optical illusions or exhalations from the Earth which caught fire in the upper atmosphere, or even the souls of great men being gloriously transported to heaven.

• • • • • • • • • • • • • •

Here They Come, There They Go:
Early Descriptions of Comets

Whatever the official explanations may have been, people still saw comets, and they recorded their observations in various ways. References to comets appear in the Bible, though in a transfigured way. In 1 Chronicles, 21:16, for example, we find "And David lifted up his eyes, and saw the angel of the Lord stand between the Earth and the Heaven, having a drawn sword in his hand stretched out over Jerusalem."

The ancient Chinese also have used the sword as a descriptive metaphor for a comet. In *The Book of the Prince of Huai-Nan*, termed "one of the most important monuments of ancient Chinese scientific thought" by Joseph Needham, the leading authority on the history of science in China, a comet apparition in the eleventh century B.C. is described this way. "King Wu marches on Zhou, faces the east and greets Jupiter, arrives at Qi, which floods, reaches Gongtou, which falls; a comet appears, giving its handle to the people of Yin." In 1979, astronomer Y. C. Chang, working at the Purple Mountain Observatory in China, announced that a computer-assisted analysis of Halley's apparitions revealed that the comet seen by King Wu in 1057 B.C. was Halley's.

The military context of this image makes us see it as a sword, but it could also have been a "broom," the more traditional Chinese image of a comet. The Chinese call comets "broom stars."

Chinese chronicles are also the source of the earliest generally accepted mention of Halley's comet, and that was in the year 240 B.C. That a comet was seen that year is practically certain, for a reference to it appears in another source almost halfway around the world.

In the summer of 1984, the official Soviet news agency, Tass, issued a scientific release based on the work of Jan Kletniek. Kletniek is on the engineering faculty of Riga Polytechnic Institute in Latvia, but he has a passionate amateur interest in old folk songs. In one of these Lettish songs he discovered an unusual line: "The sun lashed the moon with

. . . .

a silver broom." Guessing this might be a comet reference, Kletniek asked the Institute of Theoretical Astronomy at the Soviet Academy of Science for a computer check on periodic comets. According to Tass, "The computers confirmed that the tail of the now famous comet [Halley] flashed between the sun and the moon—the location specified in the song—early in the morning on May 16, 240 B.C., showing that the song-writer did not take poetic liberties."

Poetic liberties are not often taken in the Soviet Union, where social realism is the order of the day. But you can't have poetry without metaphor, and the metaphor used in the old folk song is particularly interesting because it is the same as the one used by the ancient Chinese.

Most Western observers have traditionally described the comet as a "hairy-star," comparing its tail to a flowing head of hair. The word "comet" itself is derived from the Greek *cometas*, meaning "hairy one."

To be sure, other metaphors have been used, and early de-scriptions vary according to the eye and imagination of the beholder. Of Halley's 1456 apparition, Italian astronomer Paolo Toscanelli wrote, "Its head was round and as large as the eye of an ox, and from it issued a tail fan-shaped like that of a peacock. Its tail was prodigious, for it trailed through a third of the firmament."

One of the most creative comet descriptions comes down to us from the sixteenth century, when the famous French surgeon Ambroise Paré described his view of the comet of 1528: "This comet was so horrible and so frightful and it produced such great terror in the vulgar [the common people] that some died of fear and others fell sick. It appeared to be of excessive length and was the color of blood. At the summit of it was seen the figure of a bent arm, holding in its hand a great sword, as if about to strike. On both sides of the rays of this comet were seen a great number of axes, knives, and blood colored swords among which were a great number of hideous human faces with beards and bristling hair."

The question of where a comet exists—in the Earth's at-mosphere as a meteorological phenomenon or in interplane-

. . . .

tary space as a celestial object—was not scientifically settled until an eccentric Dane, Tycho Brahe, studied the comet of 1577.

Tycho Sees a Nova

Tycho is one of the more interesting characters in the history of science. Like many other astronomers, he was turned toward his life's work by seeing a celestial spectacular, in his case a partial eclipse of the sun when he was fourteen. Also, like many other early astronomers, he was self-taught. He was the last great astronomer to work without the use of the telescope, and one of the last who still believed in the astrological significance of the stars and planets.

At the age of twenty-six, Tycho became an instant celebrity in the scientific world of Europe. On the night of November 11, 1572, as he was walking home, he happened to glance up at the sky. He couldn't believe his own eyes. Near the constellation of Cassiopeia, he saw a *new* star. It was impossible. According to Aristotle and all official authorities, the stars were unchanging and immutable. They always stayed the same.

He looked away, then back again. But it was still there. Not a comet, for it had no tail. Not a planet, for it twinkled only as stars twinkle. Perhaps he was imagining it. He quickly asked a group of peasants riding by in a wagon if they saw it too. They looked up and agreed, a "new" star had appeared.

Tycho immediately got out his instruments and began to measure the distance of his "new" star from the stars of Cassiopeia. There was no doubt it was far beyond the moon, because it remained motionless against the backdrop of the known stars. If star it was, Tycho knew that Aristotle was wrong about the nature of the unchanging stars. But he was also troubled about the astrological meaning of his find, because there were no rules to cover such an unheard-of event. Accordingly, he made his own astrological significa and reckoned that the new star's effects would be pleasant at first, then turn troublesome, and finally cause death.

• • • • • • • • • • • • • • • • •

On that night over 400 years ago, Tycho became the first methodical astronomical observer to see a nova. A nova is not a new star, it is the brief superflash of an old star exploding. Such an explosion can increase the visible light of a star from ten thousand to a million times for a few months. After this period, the star usually returns more or less to its original state. The brightness of Tycho's star would astonish any age. Astronomer C. A. Whitney of Harvard has described it as "far brighter than the brightest planets—probably bright enough to cast visible shadows on a dark night—and it could be seen in daylight." Tycho observed his star for a few months, then made his observations public. Fame was immediate.

One of the first rewards of his new fame came from the king of Denmark, who gave Tycho an island named Hveen in the Copenhagen Sound and funds enough to construct the greatest observatory in the Western world. This was before telescopes were used in astronomy, so the observatory was outfitted with complex measuring devices which resembled elaborate gun sights.

Alongside the observatory, Tycho built a private printing press, workshops, and a castle. The castle was fitting, for he was a nobleman by birth. Oddly enough, he chose a servant girl to become his common-law wife. His contemporaries were puzzled. Later, biographers offered a few explanations. Some said he married beneath his class to spite his noble peers, others that he wanted a simple wife so he could focus all his energies on astronomy. But those who seemed to know said the reason was his nose.

In his early manhood, Tycho fought a duel. He escaped with his honor, but lost most of his nose, which was sliced off by his opponent's sword. We might truly say of Tycho, he sacrificed his nose to save face.

For the rest of his life, Tycho wore a nose made of gold and silver held in place with glue. According to the nasal interpretation of Tycho's psyche, he took the servant girl because he feared no woman of noble birth would give him a second look, and if one did, she certainly wouldn't marry him.

The whereabouts of Tycho's precious nose remains a mystery. Early in the twentieth century some curious citizens of Prague, where Tycho ended his career and was buried, disinterred his well-preserved body. Alas, his gold and silver nose was missing.

In some ways, Tycho was a self-dramatizing tyrant. He dressed himself in long, specially decorated robes when he made his celestial observations, and carried on with pomp and circumstance. He treated his tenants harshly; when they were late with their rent, he threw them in jail. Eventually a new king ascended the Danish throne. He did not appreciate Tycho's mannerisms, and Tycho lost his island. The island natives soon looted and dismantled Tycho's buildings and used the stones for other structures.

Tycho and the Comet of 1577

Before his fall from royal grace, and during the heyday of his island tenure, Tycho enjoyed the contemplative sport of fishing. And it was while fishing in one of his ponds in the year 1577 that he once again noticed an unusual and brilliant object in the evening sky. At first he was perplexed, but when the sun set he saw that the object had a long, wide, glowing tail and knew at once he had discovered a comet.

Tycho studied this comet carefully during its two-month-long apparition and ascertained, without doubt, that its flight path lay far beyond the moon, somewhere out among the planets. His observations revealed that comets were celestial bodies, not phenomena occurring within Earth's atmosphere. Once again, the scientific dogma according to Aristotle was shown to be wrong.

It was Tycho's observations of the comet of 1577, and his proof that it was a celestial object, that eventually led Edmond Halley to study the comet that now bears his name.

Edmond Halley. (Royal Society/London, England.)

Chapter 2

Edmond Halley

How to Say Halley

Edmond Halley is the most famous astronomer in history, but the correct pronunciation of his last name is still a matter of minor debate, particularly among members of the Halley's Comet Society, founded in London in 1976 by Mr. Brian Harpur. What exactly the Society does is hard to pin down. Its annual meetings, carried off with British élan, seem to consist of a series of champagne toasts to the comet and its namesake plus a rousing version of the "Halley Ballad," sung to the tune of "The Battle Hymn of the Republic": "Glory, glory, Mr. Halley, your comet's hurt-ling on!" and so forth.

At the 1983 meeting of the Society, Dr. Patrick Moore tried to set the record straight: "The Halley's Comet Society has no object, no purpose, no goal, no aim, no raison d'être, and, as such, is rather like the United Nations." His remarks were greeted with applause, a "Hear! Hear!" or two, and of course a champagne toast.

The chief import of Dr. Moore's remarks, and he does wield considerable public influence as host of the BBC-TV show "The Sky at Night," was in his pronunciation of Halley. He said HAL-LEY, as in SALLY. In fairness to other opinions, though, he did recite an anonymous verse:

. . . .

"Of all the comets in the sky
There's none like Comet Halley [read HAWLEY];
We'll see it with the naked eye
And period-i-CAWley."

Halley as HAWLEY was the only pronunciation put forth
by Mr. Harpur in his remarks. To support his position, he
offered a photostat of a letter to Edmond Halley "from no less
a personage than Her Majesty Queen Anne. She spelled his
name 'H-a-w-l-e-y'."

There is at least one other pronunciation used by the British, though not nearly so popular as Halley rhyming with
Sally or Halley rhyming with Hawley, and that is Halley
rhyming with POORLY, as in HOORLY.

In America, none of the above is the pronunciation of
choice. Most Americans say Halley rhyming with Bailey or
Dailey. No doubt, much of the reason for this pronunciation
in recent years can be traced to the popularity of Bill Haley &
The Comets, who reached the top of the pop music charts in
the 1950s and left an enduring memory. Bill Haley pronounced his name to rhyme with Bailey and though the spell-
ing is different the pronunciation was applied to Halley. But
as with some other things British, many Americans are self-
conscious; anyone who mentions the man or his comet with
any air of confidence is sure to be asked, "How DO you pronounce H-a-l-l-e-y?"

The answer is simple. Anything goes; whether in poetry or
prose.

An American Credo

This is America
With freedom of speech
Where how to say last names
Is up to us each.

To the Queen's English
We don't bend our knees
So Edmond, Dear Edmond,
We'll pronounce as we please.

• • • • • • • • • • • • • • • • •

Hally, Hawley
Hailey or Hoorley
Whatever we say
We won't speak poorly.

Who Was Edmond Halley?

In his whimsical way, Brian Harpur had a point. Queen Anne did indeed instigate a letter concerning Edmond Halley, only she herself did not write it. She had her secretary of state instruct the chancellor of the exchequer of her orders: "The Queen thinking fitt to employ Mr. Edmond HAWLEY in some matters of importance to her Service abroad, has comanded me . . . that shee would have you advance him the sume of two hundred pounds towards his charges in that employment." On this occasion, Halley's assignment was to travel to the European mainland to advise the queen's ally, Emperor Leopold of Bohemia and Hungary, on the military fortification of his Adriatic ports. As in most of his undertakings, Halley was successful. In fact, his entire career was filled with successes in several diverse fields of inquiry.

Edmond Halley has been dead for 240 years, but not forgotten. His name remains the most widely known in all of astronomy. Every seventy-six years or so, the comet that bears his name is seen, or at least heard of, by almost every person on Earth. In his own day he was also famous, but not because of "his" comet. He was one of the preeminent scientists of his time, and a close friend of Sir Isaac Newton. His contemporaries could easily have imagined his name gaining immortality, but they would have had in mind his various accomplishments in such fields as theoretical mathematics, applied mathematics, astronomy, geophysics, statistical analysis, instrumentation, physical oceanography, and geomagnetism, the study of the Earth's magnetic field, which he developed.

Halley also served a brief term as deputy comptroller of the Royal Mint and was Savilian professor of geometry at Oxford University. At the age of sixty-four he was appointed Astronomer Royal by King George I.

. . . .

There was hardly a field of scientific inquiry that did not draw his attention. Some of his more obscure accomplishments, which almost seem to have been thrown off in moments of casual curiosity, would probably have been enough to secure the reputations of several scientists. One year he became intrigued with the "Bills of Mortality" for the German city of Breslau. This compendium of death statistics contained the age and sex of all persons who had died in Breslau during 1692 on a month-to-month basis. From his analysis of this data, Halley published a series of papers which provided the foundation for the study of social statistics. This work created the basic approach to the actuarial tables on which life insurance premiums are still computed to this day.

Another itch of curiosity led him to design an underwater diving bell, in which he and two other men successfully dived to depths of sixty feet, where they remained for almost two hours. An interesting feature of this device was its open bottom; water was kept out by air pressure, a design that is rudimentary but identical to the most modern underwater sea labs. He also invented the deep-sea diving suit, the mainstay of underwater work until the invention of SCUBA gear two centuries later.

Halley also made pioneering contributions in meteorology and especially in navigation, in which he devised new methods of charting the oceans, a critically important feat in light of England's move toward maritime preeminence. His chart of the trade winds and the monsoons was the first meteorological chart of its kind. His Atlantic Chart, entitled "A New and Correct Chart Shewing the Variations of the Compass in the Western and Southern Oceans as observed in ye Year 1700" enabled navigators to establish their longitudinal position while at sea. This chart, also the first of its kind, went through many editions, and remained the standard reference work for several decades.

Halley's accomplishments are so numerous that even to list them would take more space than we can allow here. But his life, truly one of the most remarkable in all of science, is so unique and so generally unknown to the public, that we do him a disservice by not telling the essentials of his story.

Edmond Halley was born in the outskirts of London in 1656. His father was a small but successful merchant, and there was no history of scholarship in the family. Edmond was an original. His lifelong romance with astronomy began at the age of ten. As he later recalled, "From my tenderest youth I gave myself over to the consideration of Astronomy" from which he received "so great pleasure as is impossible to explain to anyone who has not experienced it."

Prepared by an enormous amount of self-study and a successful career at St. Paul's School, he entered Queen's College, Oxford, at the age of seventeen. Once again, his precocious scholarship won him immediate recognition. But for a variety of reasons, including a deep philosophical difference about the nature of the universe between his thoughts and those of the established views of the Church of England, Edmond left Oxford without taking a degree.

At the age of twenty, he received what we would now call a government grant to go to the island of St. Helena, located approximately midway between the continents of South America and Africa, in order to observe the southern constellations and prepare his results for publication. An extensive, accurate, and therefore reliable charting of the stars seen from the Southern Hemisphere had never been done, and since celestial navigation was one of the mainstays of England's growing maritime power, an accurate chart of these stars was necessary for England's expanding influence in the Southern Hemisphere.

Two years later, Halley returned to England and subsequently published "A Catalogue of the Southern Stars." The quality of this work was so theoretically excellent and practically useful that his reputation was firmly established. The work was soon translated into French and German. By order of the Crown, he was awarded a Master of Arts degree from Oxford. And perhaps the greatest honor was bestowed when he was elected a member of the Royal Society. All of this by the age of twenty-two.

As a fellow of the Royal Society, and for several years its secretary, Halley was in the middle of one of the most exciting, fruitful periods in the history of science, the dawning of the Age of Enlightenment. He was in personal communication with the leading scientists of his time.

Among his keenest interests were the ideas of his friend, Isaac Newton. Not only did Halley understand these ideas, which placed him in an elite intellectual minority, but he championed them to the world. He knew that Newton's ideas were the basis of a complete revolution in the understanding of the nature of the universe. Consequently, he edited Newton's monumental *Philosophiae Naturalis Principia Mathematica*, which remains unquestionably one of the peaks of human intellectual accomplishment. Furthermore, Halley paid for the publication of this masterwork out of his own pocket. This deed alone would have sufficed to place Halley in a position of honor in the history of science.

In the *Principia*, Newton propounded the Law of Universal Gravitation and the Laws of Motion, concepts which forever altered our understanding of the physical world. Among other intellectual benefits, they provided the mathematical foundation for solving the riddles of celestial mechanics, one of which was the mysterious flight paths of comets.

Why the Comet Is Named for Halley

Newton himself believed that comets traveled in open-ended parabolic curves and did not orbit the sun. In his scenario, comets were drawn into the solar system by the sun's gravity, which bent their flight paths. They passed by once, then continued their endless flight through the universe.

Halley suspected that comets followed closed elliptical orbits, and did in fact orbit the sun. So when he turned his attention to a study of the flight paths of comets in 1695, he did so with an open mind. In his usual manner, Halley approached his new inquiry in a painstaking and methodical way. He performed two immense labors. First, he researched all the historical records he could find and collected every scrap of comet-observation data in them. Once collected, the data was sorted and everything that seemed inaccurate was discarded. Second, he meticulously computed the comet data to derive the orbital paths of specific comets. This enormous

• • • • • • • • • • • • • •

computation, done completely without the aid of mechanical computation machines, was extremely tedious and took years.

While engaged in his comet study, Halley continued his life-long habit of following many interests at once. In 1696 he became deputy controller of the Royal Mint. In 1698 he captained a scientific exploration ship under orders of the Royal Navy. In 1702, he consulted on military defense in the Adriatic Sea. In 1704, he was appointed to the chair of geometry at Oxford University.

When his comet computations were finally completed, Halley was struck by the similarity in the flight paths of three very bright comets—those which had been seen in the years 1531, 1607, and 1682. After rechecking his data and computations, he concluded that the three bright comets mentioned above were, in reality, one comet which circled the sun approximately every seventy-six years.

In the year 1705, Halley published *A Synopsis of Cometary Astronomy*, a brief work which contains the computed orbital paths of twenty-four comets. He also predicted that the bright comet last seen in 1682 would reappear at Christmas in 1758. He knew he would not be alive to see if his prediction came true, so he asked posterity "to acknowledge that this was first discovered by an Englishman."

Edmond Halley lived to the ripe age of 86 and died in the year 1742. As we've seen, his accomplishments were numerous and varied, but in his obituary notices no mention was made of "his" comet.

On December 25, Christmas Day of the year 1758 and sixteen years after Halley's death, his prediction came true. Like so many other milestones in early astronomy, this one was also reached by a self-taught amateur. On that day a German peasant-farmer from Saxony named Johan Georg Palitzch set up his homemade telescope, scanned the sky and saw the comet, just as Halley had predicted. The comet was immediately given Halley's name.

The validity of Halley's prediction served two ends. It provided the first demonstrated proof of Newton's Theory of Universal Gravitation, and it put Edmond Halley's name in lights forever.

Chapter 3

About Comets

A skyful of comets. (From The Library of Congress.)

We look up into the sky and see a stranger among the familiar pattern of the stars. It moves unhurriedly, at about the same rate as we perceive the motion of the moon. But unlike anything else in the heavens, it carries with it a long and glowing trail of ghostly light. It appears, then disappears behind the sun, then reappears once again as it begins its outward journey. It lingers in our skies for a few weeks or a few months, and then it is gone. We have witnessed the passage of one of the most fabled objects in the universe—a comet.

. . . .

What Is a Comet?

Let us begin with the mystery. Nobody knows what comets are or where they come from. Many theories have been advanced over the past few thousand years, but they remain just that—theories. The two most popular theories of modern times are the "Swarm Theory" and the "Dirty Snowball Theory."

The Swarm Theory was first proposed in the latter part of the 19th century and was still advocated by some astronomers as recently as the 1970s. According to this model, a comet was seen as a loose swarm of tiny, separate particles—a flying sandbank of sorts—that travels through space as a single unit. The particles were thought to be held together, though not necessarily touching each other, by their mutual electrical and gravitational attraction. The theory fits our perception of what the head of a comet looks like: fuzzy and apparently solid.

But the Swarm Theory does not account for some known facts about comets. For example, it is known that comets discharge and lose a good deal of their matter, in the form of various gases and small particles, when they are in the vicinity of the sun. If a comet was as insubstantial as a flying sandbank, it would lose all of its matter in a few visits to the sun. But many comets have been observed to orbit the sun more times than the Swarm Theory would allow. Halley, the best documented of all comets, will be making its thirtieth recorded visit (dating back to 240 B.C.) in 1985–1986. What's more, observed comets are not shrinking as quickly as the Swarm Theory model would predict.

The most widely held view among astronomers today is the Icy Conglomerate Model, commonly known as the Dirty Snowball Theory. This model, first proposed by astronomer Dr. Fred Whipple in 1950, presents a comet whose essential feature is a tangible, relatively small nucleus, composed mainly of a very porous mass of solidified, icy gases and solid dust particles. In the words of Dr. Whipple, the nucleus has the consistency of "a yeasty raisin bread."

. . . .

For most of the time, according to this model, the nucleus is a tiny, frozen object in space, only several miles in diameter. It only grows when it approaches the sun. In the case of Halley's comet, growth begins around the time the nucleus passes Jupiter on its return toward the sun. At this time, solar heat begins to melt the various ices. In actuality, they don't melt, they sublimate, which means they turn directly from solid ice to gas, with no liquid stage between. During sublimation, the outer layer of the nucleus loses its shape and some of its material. A vapor envelope, composed of the escaping gases and dust, begins to create the comet's head. As the head nears the sun, more ices sublimate and the head continues to grow. Within a certain range, the gases and dust particles are caught up by the solar wind, a powerful force emanating from the sun. Some of these gases and particles are blown by the solar wind and form the comet tail we see from Earth.

Once the comet circles the sun and begins to depart, sublimation stops. The head and the tail appear to shrink. This is because the particles are dissipating into space and are no longer being replenished by further sublimation.

Because the nucleus is such a poor conductor of heat, only an outer skin of material is lost on each orbit. The heart of the nucleus remains frozen solid. Most comets are capable of making hundreds, perhaps thousands of orbits around the sun before their mass diminishes enough to allow them to be swept into a new orbit.

Where Do Comets Come From?

The origin of comets remains unknown, though several theories have been offered by astronomers. Some say that comets are fragments of other celestial bodies that have exploded. Others say they are formed when interstellar gas clouds condense or collapse. The most interesting theory, and one that has received the most attention in recent years, is based on the Oort Cloud.

The Oort Cloud, (actually first proposed in the 1930s by

• • • • • • • • • • • • • • •

Estonian astronomer Ernst Opik) is named for its major modern advocate, astronomer Jan Oort of Holland. According to this theory, the entire solar system is encircled by a vast ring of comets, meteors and smaller particles. The inner edge of the Oort Cloud is hypothetically placed beyond Pluto, the furthest known planet in the solar system. The outer edge of the cloud is thought to extend to the outermost fringe of the sun's influence. The scope of this cloud is staggering: it is estimated to be about 10 trillion miles in diameter.

According to this theory, the Oort Cloud is where comets are born. It is the "comet nursery" of the solar system. The smaller particles of matter are coalesced into comet nuclei by the gravitational effects of a passing star or one of the larger planets when they pass close to the edge of the cloud. Moreover, these gravitational effects are thought to provide the nudge that comets need to break free from the cloud. Many of the dislodged comets are thought to take off on flights through interstellar space. But some fraction of the comets born in the Oort Cloud are drawn into the solar system. These are the comets that eventually visit the sun.

The origin of the Oort Cloud is thought to be the Big Bang, the mysterious event that gave rise to the entire universe. If this is true, the matter that forms the cloud is in an almost pristine state; it is matter that has changed very little since the Big Bang. Since it is the stuff of which comets are made, a closer study of comets could very well reveal part of the answer to the biggest question of all: What is the origin of the universe?

The Parts of a Comet

- The Nucleus: This is the small, frozen heart of a comet, the enduring object that actually makes the full orbital swing around the sun.
- The Head, or Coma: This is the visible comet we see in the sky. It is formed by the sublimated gases and small particles emitted by the nucleus when it is within the sun's "melting" range. The gases and particles are held

. . . .

"close" to the nucleus by gravity and electrical attraction. The head of a comet can grow to be 5 million miles across.

- The Tail: Comets have two types of tails. They are created by different forces and they look quite different to us here on Earth. One is the *dust tail* and the other is the *ion tail*.

The *dust tail* is composed of tiny dust particles, which are about one micrometer (one-millionth of a meter) wide, and of ice particles, also very small. Such tails are among the most spectacular sights in the universe. Some have grown to more than 100 million miles in length and 50 million miles in width. Their volume can exceed 50 trillion cubic miles. But when we look at a comet's tail, there's less than meets the eye. It is so insubstantial that each cubic yard of its volume contains only *one* molecule of solid matter or gas. If we could gather up such a tail and put it on the scale, it would weigh in at only *one-half of one ounce!*

Scientists currently think that the dust is largely silicate, or sand, with an unknown amount of carbon mixed in. As these particles reflect sunlight, they become visible and we see them, not as individual specks, but as a sweeping tail of light.

The *dust tail* is driven away from the comet by the solar wind, and *always points away from the sun*. In a sense, the term "comet tail" is a misnomer, since we tend to think of tails as appendages that trail *behind* their owners. A comet's tail does not necessarily trail behind. Only as a comet approaches the sun does the tail wag behind. Most of the time, the tail points downward, at a right angle to the comet's path. And as a comet leaves the sun and heads for the other end of its orbit, the tail precedes the head.

Our perception of a comet's dust tail is altered in another way. Because this tail is held by the gravitational pull of the comet, it travels the same orbit, and thus it often appears to be curved. The curvature is actually the curve of the comet's orbit.

The *ion tail* is more mysterious. As molecules of matter are

. . . .

blown off and away from the nucleus, they become ionized
by the sun's ultraviolet and X-ray radiation. Ionization means
that an electrically neutral atom takes on an electrical charge.
In a manner similar to the way a fluorescent light bulb glows,
the sun's electromagnetic radiation causes the ions to glow.

Because the *ion tail* is largely electrical, rather than physical, it does not appear to curve along the orbit of the comet.
Instead, it appears to point almost straight out, directly away
from the sun. About every week or so, however, the *ion tail*
is caught off guard when the polarity of the solar wind
changes. Suddenly, the particles that had attracted begin to
repel, and those that repelled begin to attract. The result is
that the visible *ion tail* actually disconnects from the head of
the comet. The ions regain an equilibrium in about thirty
minutes, however, and the *ion tail* rejoins the comet's head.

The complexity of a comet's tail is compounded by an item
called an *antitail*, which frequently points *toward* the sun.
Antitails are not "real" tails. Scientists, in fact, refer to them
as "projection effects." As the comet orbits, it leaves a trail of
dust behind. This trail has nothing to do with melting ices or
the solar wind, it is merely a trail of comet debris. In just the
right position and at just the right time, this trail reflects light
from the sun. From our vantage point here on Earth, the result
of this effect is a comet with tails pointing in two directions.

The trail of comet debris does not end with the antitail, nor
does it disappear. It remains where it is and provides us with
occasional displays of celestial pyrotechnics. Meteor showers
are caused by Earth passing close to a comet's orbit, even after
the comet is long gone. The particles of debris are drawn into
Earth's gravitational field and rush toward us at extreme ve-
locities. As they pass through our atmosphere, friction burns
them up and we see them as shooting stars. Halley's comet,
or at least the trail of debris left along its orbit, shows itself
twice every year: the Aquarid meteor showers of May and the
Orionid showers of October are delightful gifts from Halley to
us.

The mysterious behavior of comets' tails probably helps to
explain some of the dread that comets create on Earth.

Why Halley?

Of all comets, why is Halley the most famous, to scientists and laymen alike? One reason is the span of its known history . . . we have observed its traveling pattern for over 2000 years, far longer than any other comet. Another reason is Halley's makeup; it possesses nearly all of the significant features of a comet, including mysteries. One is the fact that despite its age, estimated to be in the range of 4000 to 5000 years (on its present orbit, this would mean it has circled the sun between fifty and sixty-five times), the rate at which it releases gas and dust resembles that of a younger comet.

Halley's close passage to Earth (in the range of 50 million miles) enhances terrestrial observation and, for the 1985–1986 visit, makes space exploration feasible. Its close passage to the sun (also in the 50 million mile range) makes it a very active comet. It is big; it is bright; it has a large coma; it has developed both dust and ion tails. And of greatest importance to most of us, it has always been visible from Earth.

The only drawback Halley presents to scientific investigation is the direction of its travel. Earth, the rest of the planets, and most other comets orbit the sun in a counterclockwise direction. But Halley's orbit is clockwise, what astronomers call "retrograde." As a result of this retrograde orbit, Halley and Earth, traveling in opposite directions, pass each other at exceptionally high speed. Though Halley will be visible to most of us for a few months and to astronomers for a few years, close-up observation presents a problem. The closest space probes being sent to rendezvous with Halley will pass and photograph the comet at speeds seventy times faster than a bullet!

During Halley's 1985–1986 visit, its first in the Space Age, we shall learn more about comets than all the knowledge gathered since our earliest ancestors first looked up and saw a glowing stranger float across the familiar constellations. When Halley is in view, spacecraft will ascend from several locations on Earth and approach the comet to gather data.

Much of our current "knowledge" of comets is theoretical.

It is the hope of scientists that the comet probes during the visit will provide hard answers to many of their questions. For example, they hope to gain much more precise information about the size, the weight, and the composition of Halley.

How Big Is Halley's Comet?

The size of the nucleus of Halley's comet is a matter of some dispute. A scientist at the Goddard Space Center has estimated that it is three miles in diameter. Carl Sagan thinks the diameter is about 12½ miles. More recently, scientists at NASA's Jet Propulsion Laboratory (JPL) have also estimated that the diameter is probably about three miles, though they hedge their estimate by allowing for a plus or minus error ranging from half a mile to ten miles.

How Heavy Is Halley's Comet?

Since the composition of a comet is still unknown, it is difficult to estimate its weight. The weight and size of a comet are numbers people like to know when they come to fantasize about a collision between Earth and a comet. Working backward from what is known about Halley's orbit, and how it is affected by the gravitational pull of various planets, the JPL scientists have estimated Halley's weight to be at least 65 billion tons. This sounds very heavy, but comparatively speaking, it is a featherweight. Our Earth, for example, weighs 6.6 sextillion tons (that is, 6.6 followed by 21 zeros) and is 100 billion times heavier than the estimated weight of Halley. An equivalent match-up might be a mosquito and Moby Dick.

What Is Halley's Flight Path?

Its relative lightness among the planets of the solar system makes Halley extremely vulnerable to numerous gravitational pulls, particularly those exerted by the giant outer planets: Jupiter, Saturn, Neptune, and Uranus. Depending on where they happen to be in their own orbits around the sun when

· · · · · · · · · · · · · · · · · ·

Halley passes by, they exert a greater or lesser pull. The celestial mechanics involved are highly complex, but the advent of computers has helped astronomers to make these calculations more quickly and probably more accurately. Nevertheless, it remains difficult to make precise predictions about a comet's arrival time, which is usually reckoned at perihelion, its closest approach to the sun.

The effects of gravity vary according to distance. Halley, like all comets, travels more quickly when it is close to the sun. At perihelion, Halley zips along at approximately 120,000 miles per hour. At aphelion, when it is furthest from the sun, it plods along at a mere 2000 mph.

Once Halley arrives, its physical behavior from day to day will also defy prediction. There is simply no way yet to tell how long its tail will be, or what shape, or how brightly it will glow in the heavens. As one of the JPL scientists has warned, "Anyone who attempts to predict the physical behavior of an active comet is almost certain to be incorrect." The physical behavior referred to has to do with the comet's tail, which can stretch halfway across the sky and be bright enough to cast shadows.

When it comes to the question of flight path, comets fall into two major categories: those that orbit the sun, and those that don't. A comet that does not orbit the sun is a one-time visitor to our solar system. Its flight path is bent by the combined gravitational pull of the sun and the planets, and it veers in toward the sun, but it is not captured. After its solar flyby, it continues off into space.

Orbiting comets are controlled by the gravitational pull of the solar system. They fall into two subcategories: long term and short term. Long-term comets have immense orbits which carry them far from the solar system, but not far enough to be captured by the gravitational pull of another star. Their orbiting cycle is measured anywhere between hundreds and thousands of years.

Short-term comets travel in orbits which hold them within the solar system. Some, the so-called Sun Grazers, complete their orbits in a year or two. Halley has the longest cycle of all the short term comets. Every seventy-six years, on average, Halley passes through our skies.

Halley's timetable is not precise because its orbit is affected by the gravitational pull of other celestial objects. Depending on how close it travels to any or all of the planets, whose gravitational pull can slow it down, its return time can vary by as much as five years. Halley's shortest return on record was 74.4 years (between 1835 and 1910) and its longest was 79.2 years (between 451 and 530 A.D.).

After Halley circled the sun in 1910, it turned and headed out toward the other end of its elliptical orbit, which lies somewhere between Neptune and Pluto, the two farthest known planets in the solar system. When it reached that distant point in 1948, the sun's gravitational pull turned it around and caused it to head back in toward the center. It has been on its way back to us since then.

Halley's comet is undoubtedly Earth's most famous visitor from the depths of the solar system. It will return to our skies in the fall of 1985 and again in the early spring of 1986, after it has circled the sun. This will be Halley's first visit to Earth since 1910, but it will not arrive as a stranger. Dr. Brian Marsden, director of the Harvard Smithsonian Astrophysical Laboratory, has remarked, "To the man in the street, the solar system consists of Mars, the Rings of Saturn, and Halley's Comet."

Chapter 4

Comet Hunting

In the language of astronomers, a visit by a comet is called an "apparition." It is an interesting word choice, echoing as it does the aura of mystery that surrounds comets. In common word usage, an apparition is "an unusual sight" or a "ghostly figure," terms which describe how comets are perceived by most people.

Halley's 1910 apparition had been expected for seventy-five years, since its previous appearance in 1835. Astronomers had a reasonably good idea when it would enter the range of the most powerful telescopes then in existence. And since they knew the basic path of its orbit, they also knew in which direction to look. In the winter of 1908–1909, the search to "recapture" Halley's comet began in earnest.

Until that time, the search for comets had been carried out by astronomers peering through telescopes. Now a new technology was added to the search for Halley—photography. Specially designed cameras were attached to telescopes, the telescopes were aimed in the direction of Halley's expected arrival, and photographic plates were given long exposures. When the plates were developed, they were carefully scrutinized for any object which appeared to be in motion; this

The Search Begins. (From The Library of Congress.)

. . . .

would make the object show up as a tiny streak or blur. Since stars, so far as we know, maintain a fixed position relative to each other, any object that moves against the background of fixed stars would have to be a nonstar, or in this case, a comet.

This new photographic technique was a great improvement, and it remains in use to this day, employing more sophisticated cameras and more powerful telescopes. This photographic method has been the way that all returning comets have been recovered since 1910. (For its 1985–1986 apparition, Halley was photographically recaptured by astronomers at the California Institute of Technology on October 16, 1982, when it was about a *billion* miles from Earth and three years and four months from perihelion, by far the earliest recapture of a comet in history. The Cal Tech astronomers used an advanced electronic detector system, designed for the space telescope scheduled for launch in 1986, and the 200-inch [diameter of the main lens] Hale telescope at Palomar Observatory. By comparison, the recapture of Halley in 1909 took place only seven months before its 1910 perihelion. And as a further example of the progress that has been made in the design of telescopes and in the plotting of Halley's orbit, an *amateur* astronomer named Tsutomu Seki from Kochi, Japan, photographed Halley in late September, 1984, over sixteen months before perihelion.)

If the sky camera increased the quality of celestial observations, it also increased the traditional competition among skywatchers. Astronomers, professional and amateur, have long vied for the distinction of discovering new comets or for being the first to spot a returnee.

Since the mid-eighteenth century, when comet hunting began to flourish, fewer than 550 comets have been discovered, so the odds against discovering a new one are long. But the reward is great—a kind of immortality. A new comet is given the name of its discoverer. If a new comet is discovered simultaneously by two or more observers, it receives a multiple, hyphenated name, like the tongue-twisting Whipple-Fedtke-Tevzadze, the staccato Tago-Sato-Kosaka, or the more melifluous Swassmann-Wachmann.

. . . .

For the discoverer of a new comet, or for the recoverer of a known comet, accuracy and speed are essential. The credit and honor for a new sighting go to the observer who makes the first formal and eventually verifiable claim. An observer stakes his or her claim by sending a telegram to the telegraph bureau of the International Astronomical Union (IAU), preferably through the national observatory of the observer's own country. Since the early 1960s, the bureau has been located at the Harvard Smithsonian Astrophysical Observatory in Cambridge, Massachusetts.

The details of the observer's claim (what was seen, where and when) are checked at the observatory, then quickly sent to astronomers around the world for added verification. If all goes well, the observer's claim is formalized and, in the case of a comet discovery, his or her name is forever tied to the new comet's tail. The "his or her" designation is not used hypothetically. In a thirteen-year period ending in 1959, Ludmilla Pajdusakova and her staff at the Skalnate Pleso Observatory in Czechoslovakia discovered eighteen new comets, one of the most productive runs by any comet-hunting team in history.

The IAU did not yet exist in 1910—it was originally organized in the 1920s—but there was a generally accepted verification process subscribed to by astronomers around the world. Basically, an observer who believed he had sighted a new comet or recaptured a known one would make a public announcement, inform his colleagues, and get the word out. Other observers would respond to the information and the verification process would unfold. The public, as always, was ready and eager to hear the news.

Recapturing Halley in 1910

The search to recapture Halley began at the Yerkes Observatory in Wisconsin, home of a 40-inch refractor telescope, which was, and remains to this day, the largest of its type ever built. The Yerkes staff, which included Edward Barnard, a world-class comet hunter, exposed its first photographic patrol plate late in December of 1908. But it was not until nine months later that Halley was recaptured. The honor of being the first person to recapture a comet by use of photographic plate went to Dr. Max Wolf of Heidelberg, Germany, then a major center of astronomical studies. Upon careful scrutiny, Wolf discovered a minute image of Halley on a photographic plate he'd exposed on the night of September 11, 1909. It's not surprising the image was minute. On that date, Halley was 331 million miles from Earth.

Following Wolf's announcement, other observers went back and rechecked their early plates. As it turned out, Halley had in fact been photographed before September 11, but its image had originally been missed. The earliest photo image, taken on August 24, 1909, was obtained at the Helwan Observatory, located opposite the ruins of ancient Memphis along the Nile River in Egypt.

Meanwhile, back at the Yerkes Observatory in Wisconsin, the search shifted from camera to eye. On September 15, S. W. Burnham, peering through the observatory's 40-inch refractor telescope, became the first person to actually see Halley's comet since May 1836, when it was last seen by Sir John Herschel through his 4-inch refracting telescope. By an odd coincidence, Herschel's small telescope was in use as the guiding telescope at the Helwan Observatory when its large photographic reflector telescope caught the first photo image of Halley in 1909.

Edward Barnard and the House Built of Comets

Burnham, whose specialty was double stars, had little interest in comets. But two nights later one of the most famous and successful of all comet hunters, Edward Barnard, had his scheduled turn at the 40-inch telescope and he too saw Halley. It was entirely fitting that Barnard should have been one of the first people to see Halley on its 1910 return. During his long career, he discovered sixteen comets and the fifth moon of Jupiter. He is also one of the most interesting of comet hunters.

· · · · · · · · · · · · · · · ·

Barnard, surely one of the most endearing of all astronomers, began his comet hunting career as a self-taught amateur.
His searches were so successful, and his fame spread so far
though the astronomical community, that he was offered a
faculty position in practical astronomy at Vanderbilt University. Later he served as a professional astronomer at the Lick
Observatory, and finally at Yerkes. In the truest sense of the
word, Barnard was a cometophile. To be sure, he made many
contributions to several fields of astronomical studies, including pioneering work on the use of sky cameras. But comets
were his great celestial love. Barnard spent most evenings of
his life looking for them. He even dreamed about them. One
of his comet dreams, which happened while he was taking a
brief nap during a night of comet searching, was portentious.

"I had the most wonderful dream," Barnard once recalled.
"I thought I was looking at the sky which was filled with
comets, long tailed and short tailed and with no tails at all . . .
I had just begun to gather the crop when the alarm went off."
What happened next was more amazing than the dream itself.
Barnard got up, took his small telescope out into his back-
yard, and began to sweep the sky. Almost immediately he
saw not one comet, but a swarm of them. "Before dawn killed
them out I located six or eight of them."

Barnard sent a telegram to a leading astronomer, but for
some reason his sighting information was not forwarded to
other observers and he received no credit for this observation,
even though it was verified by other observers in the United
States and Europe. Barnard believed the swarm of comets he
saw that night were actually fragments of a great comet that
had recently grazed the sun. As for his unusual prevision, he
later remarked, "The association of this dream with the reality has always seemed a strange thing to me."

Barnard's involvement was so complete that he even built
his house of comets. This novel building technique came
about like this: In 1880, a wealthy American businessman
named H. H. Warner offered a $200 cash prize to the discov-
erer of every new comet. In those years, when the average
annual wage for an American worker was about $700, $200

was a significant sum of money. So when Barnard won such a prize in 1881 for his second comet discovery, he invested in a plot of land, took out a mortgage, and built a small house for his wife, his mother, and himself. Since he spent most of his time looking for comets, the problem was how to pay the mortgage. As he recalled years later, he looked "forward with dread to the meeting of the notes that must come due. However, the hand of Providence seemed to hover over our heads; for when the first note came due a faint comet was discovered . . . and the money [from the Warner Prize] went to meet the payments." From then on, Barnard often managed to discover a new comet just in time to pay off one of the "dreadful notes."

Looking back on that period of his life, he expressed great satisfaction. "And thus it finally came about that this house was built entirely out of comets." And with a touch of whimsy, he added, "This fact goes to prove further the great error of those scientific men who figure that a comet is but a flimsy affair after all; for here was a strong, compact house— albeit a small one—built entirely of them. True, it took several good-sized comets to do it; but it was done nevertheless!"

On February 10, 1910, Barnard measured Halley's tail: it was then about 5 million miles long. Astronomers had recently announced the discovery that Earth was heading for an encounter with the comet's tail, but Barnard, like the majority of his colleagues, was not worried. Instead, he told the public "there was great reason to hope that by May 18 there would be plenty of tail to reach to and beyond the Earth."

Barnard was a charming man, and one of the world's leading cometologists, but his understanding of the comet's power over the human imagination was naive. While he and his colleagues awaited Halley with eagerness, the mass of people felt a growing sense of apprehension and dread.

With its attendant figure of Death, the comet wreaks havoc. (From the Zwerdling–Miranda Collection.)

Chapter 5

The Curse of
Halley's Comet

W hen Edmond Halley's name was pinned to the comet whose orbit he discovered, he became forever associated with one of the most interesting physical objects in the solar system. He also became associated with an important scientific discovery, and, on another level of human behavior, with the very ancient tradition of cometophobia, or fear of comets.

A deep fear of comets seems to have been lurking in the background of our imaginations forever. The appearance of a comet traditionally provokes a widespread fear that some kind of disaster is near at hand. The very word itself, "disaster," derives from *dis* (evil) and *astrum* (star), and some experts have traced the "evil star" to comets.

Throughout history, numerous attempts have been made to compile a "record" of the disasters associated with the visits of comets. One of the most interesting examples of these records was compiled by Increase Mather.

The Reverend Increase Mather, pastor of North Church and president of Harvard College, was one of the intellectual and moral leaders of Puritan New England. He was so concerned about the destructive powers of comets that he researched

• • • • • • • • • • • • • • • •

ancient records and in 1683 published his findings. The title tells the story, and is representative of similar efforts made by writers throughout the centuries: *A Discourse Concerning Comets In Which Their Origin and True Matters Are Enquired into, and The Most Remarkable Earthquakes, Famines & Wars Attending Their Appearance from the Beginning of the World, Chronologically Arranged.*

Any comet can cause cometophobia, but the most ominous comet of all is Halley's. Since the discovery of its orbital path, its previous apparitions have been traced backward through history. Perhaps the best-known example of Halley's Curse occurred around the apparition of 1066.

Poor King Harold

The year 1066 is one of the most memorable in history, certainly to long generations of English schoolchildren who have had to memorize it. That was the year that William, Duke of Normandy, crossed the English channel and defeated King Harold at the Battle of Hastings. The duke lives in history as King William the Conqueror, the event lives as the Norman Conquest.

As we have been told, "History is written by the victors." In this case, it was celebrated and illustrated in a remarkable document known as the Bayeux Tapestry. Actually, it is not a tapestry at all; it is a crewel embroidery, executed in eight colors, on coarse linen.

Tradition has it that William's wife, Queen Matilda, and a flock of her handmaidens created the embroidery. The reality is a little less romantic. Matilda, eager to commemorate her husband's great triumph, commissioned a group of artisans to create a pictorial record of the great event. The embroidery was ten years in the making.

The length of time it took to complete the piece is understandable when we consider its size: it is 19½ inches wide and 231 *feet* long. In its totality, it is a unique pictorial record of the history and costume of the period. Currently it hangs in the Town Hall of Bayeux in Normandy, France.

• • • • • • • • • • • • • • • • • •

What is of most interest to us in the embroidery is the depiction of the loser, King Harold. The ill-fated king sits on his throne, his head tilted to the side as he listens to the words of a messenger. Harold's expression is glum, and for good reason. In all likelihood, the messenger is delivering the ominous news that the Norman army has just landed on the English coast.

The viewer's eye is next carried to the upper margin of the panel. Above and to the left of Harold, cruising mysteriously over the scene, is nothing other than Halley's comet, that great augury of disaster.

In the next panel, a group of Harold's Saxon subjects, also looking disheartened and fearful, point up at the comet. The embroidered Latin caption says, *"ISTIMIRANT STELLA,"* which means "They are in awe of the star." This is the earliest documented contemporary representation of Halley, and it remains one of the most interesting.

Halley's 1066 apparition was also recorded in *The Anglo Saxon Chronicle:* "In this year King Harold came from York to Westminster at Easter . . . Then was seen over all England such a sign in the heavens as no man ever before saw; some say it was the star Cometa, which some men call the haired star; and it first appeared on . . . the 8th of the Kalends of May [April 24th] . . ."

The "sign in the Heavens such as no men ever before saw" was generally understood to be an omen of disaster. There was little doubt that comets had a direct impact on human fate. Halley's 1066 visit occurred in March, the Battle of Hastings in mid-October, seven months later. But there has never been a time limit on superstitions, no precise schedule for how soon the effect must follow the cause. So when Harold fell at Hastings, his death was taken as an undisputed proof of the danger of comets. And William's capture of the English throne neatly demonstrated a proverb of the day, which said that a "new star" (the comet) meant a new king.

Little if any mention was made of another great battle that occurred in 1066. This one happened in September, even closer to Halley's appearance, and it also included King Har-

• • • • • • • • • • • • • • • •

old. At the Battle of Stamford Bridge in northern England, Harold's Saxon army successfully repelled an invasion and slew the two enemy leaders, the king of Norway and his ally, a pretender to the English throne named Tostig. This victory would seem to cancel out the defeat at Hastings and the effect of Halley's curse. It also provides a rational reason for Harold's poor showing at Hastings: he and his army were exhausted. There was also a technical explanation: the Norman army was better trained and better equipped. But for those who believe in Halley's curse, such details are only coincidental trivia compared to the power of the comet.

Advancements in the scientific understanding of comets have not eliminated the superstitious dread with which many people view comets, especially Halley's. From the vantage point of our super-sophisticated-high-tech-computerized-nuclear space age, we tend to assume that superstition is on the wane. But there is another stream in contemporary life, a kind of counter-current: throughout the world today, there are numerically more adherents of astrology, occultism, and other supernatural systems than ever before. Rationalism and nonrationalism are still in competition. The fact is, there is no way to tell how the world will respond to Halley during its 1985–1986 apparition. We have no way of knowing how many comet cults will appear to announce the end of the world, or how many people with an already tenuous grasp on reality will be driven to let go altogether.

The enlightened rationalists of 1910 also believed superstition was on the wane. As the world's astronomers planned how they would study Halley, and began to look for it in the heavens, they had no idea of how intensely, even obsessively, interested people soon would be in their every observation and opinion.

Cover of Scientific American: U.S.A. A comet watcher's guide, with
the constellations above and the Manhattan skyline below. (From
The Library of Congress.)

DISCOVERY OF A COMET AT GREENWICH OBSERVATORY

Above: *Drawing by Henri Lanos from* The Graphic: *England. Hot-air balloons were as close as astronomers could get in 1910. This time around, space probes will approach to within 350 miles of Halley's comet. (From The Library of Congress.)*

Opposite, top: *Drawing by Georges Plasse from* L'Illustration: *France. A member of a French astronomical expedition to North Africa views the comet. (From The Library of Congress.)*

Opposite, bottom: *Cartoon by G. Morrow from* Punch: *England. Comets are unrivaled when it comes to attracting telescopes. (From The Library of Congress.)*

L'ILLUSTRAZIONE ITALIANA

Anno XXXVII - N. 21 - 22 Maggio 1910.

Per tutti gli articoli e i disegni è riservata la proprietà letteraria e artistica, secondo le leggi e i trattati internazionali.

LA COMETA DI HALLEY.

Drawing by Rodolfo Paoletti from L'Illustrazione Italiana: Italy. Viewing the comet from Brera, Italy. (From The Library of Congress.)

How Astronomers have Seen Halley's Comet.

Drawing by C. W. Wyllie from The Sphere: England. An illustration of the comet and its tail as seen through a telescope. (From the Library of Congress.)

Drawing by S. N. Gros-nitskii from Izviestiia Russkago Astronomi-cheskago Obshchestva: Russia. The comet as seen from Armavir in the Caucasus. (From The Library of Congress.)

Drawing by Fernand Baldet from L'Illus-tration: France. The comet seen over the bridges of Paris. (From The Library of Congress.)

Postcard: France. "The End of the World, May 19, 1910. Expeditions to the Moon. High Speed! 200 francs a shot. The last salute! Goodbye and thanks!!" The sign on the dirigible reads, "We aren't doing too badly." The Moon says, "You are all very welcome." (From the M. Zwerdling–D. Miranda Postcard Collection.)

Postcard: France. "Excursions to the Moon on the Clement-Bayard II. Last flight today at 10:30! Price—2 francs per person." Sign at left: "Public access to the dirigibles." Title at lower right: "The last moments before the End of the World." Another example of a comical escape fantasy. (From the M. Zwerdling–D. Miranda Postcard Collection.)

Opposite, top: *Ad for Sapolio from* Canadian Magazine: *Canada.* (From the Library of Congress.)

Opposite, bottom: *Ad for Newmark's coffee from* The Los Angeles Times Illustrated Magazine: *U.S.A.* (From the Library of Congress.)

Below: *Ad for· Bird's custard from* Sphere: *England.* (From The Library of Congress.)

Ad for The Illustrated London News from The Sketch: England. (From The Library of Congress.)

Ad for Muebles (Furniture) Thompson from Caras y Caretas: Argentina. "A messenger from the infinite (the comet) brings glad tidings from Nature, even to the inhabitants of the most remote areas, on the Centennial of our country." (From The Library of Congress.)

STAYED UP TO SEE IT.

NOCTURNAL WANDERER.—Orfsher, 'blige me! Whish one er thosh ish Halleysh Comet?

Cartoon by L. M. Glackens from Punch: England. A comet version of the comical drunk. (From The Library of Congress.)

HALLEY'S COMET.

WILLIAM II.—"The end of the world? Impossible! I have given no such order." —*Pasquino* (Turin).

Cartoon from an Italian publication, reprinted in The Literary Digest: Italy/U.S.A. A jab at the dictatorial image of Kaiser Wilhelm. (From The Library of Congress.)

Cartoon by W. H. McDougall from The Literary Digest: U.S.A. The stout President Taft looks at the comet, but he sees the stylized face of his predecessor, Teddy Roosevelt. It was an ominous vision. Two years later Roosevelt challenged Taft, split the Republican party, and brought about the election of Democrat Woodrow Wilson. (From The Library of Congress.)

THE COMET IS NOW VISIBLE AT WASHINGTON.
—McDougall in the Philadelphia *Telegraph*.

Cartoon by R. Rost from Jugend (Munich), reprinted in The Literary Digest; Germany/ U.S.A. The czar of Russia sees the comet and assumes it's another revolutionary bomb. (From The Library of Congress.)

A "FAN'S" FANCY

Cartoon from Life: U.S.A. (From The Library of Congress.)

Cartoon from Fliegende Blaetter: Germany. "A Munich Kid." Telescope man: "See the comet, only 20¢!" Father: "Do you want to see it, Karl?" Little Karl replies: "C'mon, Pop, shell out another 8¢ and we can each have a beer instead!" (From The Library of Congress.)

Drawing by R. O. Yardley from Harper's Weekly: U.S.A. A couple on a New York rooftop find romance more interesting than the comet, which hovers over the city's skyline behind them. (From The Library of Congress.)

Cartoon from Pasquino (Turin) reprinted in Le Rire: Italy/France. "The end of the world did not arrive. Evidently, it [the comet] was only a rabbit." (From The Library of Congress.)

LA FIN DU MONDE N'EST PAS VENUE

— Évidemment, c'est un lapin !   (*Pasquino*, Turin.)

WAITING FOR THE END OF THE WORLD

Drawing by J. Hill from Harper's Weekly: U.S.A. "Waiting for the
end of the world." (From The Library of Congress.)

Part Two

Halley Meets 1910

Read All About It! (From The Library of Congress.)

Chapter 6

The Halley Drumbeat Starts

The night of May 18–19, 1910, was rapidly being shaped into one of the most extraordinary nights in history. On that night the vast majority of the human race participated in an unprecedented homage to Halley's comet.

The world's response to Halley's 1910 visit was shaped by several forces. Underlying everything was the instinctive fear of comets that has always been a human trait. All a comet has to do to cause comet fever is to make an appearance, even if unannounced. But this was Halley's, the most famous and dreaded of all comets, and it had been on its way back to us for seventy-five years.

The unique intensity of the 1910 epidemic of comet fever grew directly out of two scientific discoveries. One, based on refined calculations of the orbital paths of Earth and Halley's comet, revealed that Earth would pass through the comet's tail on the night of May 18. The other discovery was about the nature of the tail itself. A spectrographic analysis of the light emitted by the tail revealed the presence of cyanogen gas, a deadly poison.

The coming encounter with Halley's tail received as much coverage in the international press as any natural phenome-

• • • • • • • • • • • • • •

non in history. As we shall see in the following pages, much of that press coverage, intentionally or unintentionally, seemed designed to raise anxieties all over the world.

The comet stories fell into a few main categories: interviews with astronomers; background articles about Edmond Halley and comets in general; and an endless stream of pieces about "The Curse of Halley's Comet."

All of the comet curse stories covered the same body of material, namely the long list of Halley's past apparitions and the disasters that had accompanied them. The major differences in the comet curse stories were in the different ways they were presented by different publications. Some of the stories were tongue-in-cheek, and teased their readers to "believe" such superstitious nonsense. The more sensationalistic newspapers and magazines presented the very same material with breathless alarms and warnings. But the majority of publications straddled the two extremes, and consequently delivered mixed messages.*

Death, Doom, and Defeat

The earliest reported apparition of Halley's comet that is generally regarded as accurate occurred in 240 B.C. Halley's visit of that year was described and recorded in an ancient Chinese chronicle. In 1910, however, soothsayers, cometophobes, and many newspapers were less demanding in their requirements for documentation, particularly when it came to listing ex-

---

*Perhaps there is no responsible way to avoid the mixed message when Halley is concerned, because the mere mention of the curse is enough to open the door to superstitious speculation. On Sunday, December 16, 1984, the *New York Times* published a front page article: "Soviet Spacecraft Heads for Comet." It was about the launching of Vega 1, which is scheduled to reach Halley In March 1986. (See chapter 18 for a fuller discussion of space missions to the comet.) The article concerns the technological advances in the study of cometology and the various spaceship explorations that are planned for Halley's current visit. Nevertheless, the second paragraph of the story carries the subhead "A Legend and an Omen" though the copy itself merely informs the reader that "Halley's is legendary as an omen of ill-fortune," and then returns to the details of the Vega mission.

• • • • • • • • • • • • • • •

amples of Halley's curse in action. The following list of disasters linked with Halley is a composite drawn from numerous 1910 publications. The historical accuracy of these "comet disasters" is less important than the fear their publication instilled in millions of readers.

2616 B.C.   *The death of Methuselah (according to tradition).*

2349 B.C.   *The Biblical Flood (tradition).*

1900 B.C.   *The destruction of Sodom and Gomorrah (tradition).*

240 B.C.   *War between Rome and Carthage.*

66 A.D.   *Roman Emperor Nero, warned of the comet's curse, executes all of his likely successors.*

218 A.D.   *The death of Emperor Macrinus*

295 A.D.   *Bloody persecution of the Christians under Emperor Diocletian.*

374 A.D.   *Start of the rebellion in the African provinces which leads to the breakup of the Roman Empire.*

451 A.D.   *Attila the Hun, "Scourge of God," is defeated at the Battle of Chalons, changing the course of history.*

530 A.D.   *The death of King Arthur (tradition).*

607 A.D.   *Mohammed has the holy visions that will lead to the establishment of the Islamic religion.*

684 A.D.   *The Saracens conquer North Africa.*

837 A.D.   *The Danes ravage England; the Saracens conquer Sicily.*

912 A.D.   *War in Europe between the Roman Emperor and the Saracens.*

1066 A.D.   *The Turks conquer Egypt; the Normans conquer England.*

1222 A.D.   *Europe ravaged by the war between the Guelphs and Ghibellines; Genghis Khan is in the midst of conquering Asia.*

1456 A.D.   *Halley's comet exorcised by Pope Calixtus. (This episode was embellished into the popular legend*

. . . .

that Calixtus excommunicated the comet.) Austria is wracked by civil war; Janos Hunyadi defeats the Turks at the Battle of Belgrade, stopping the westward conquest of the Turkish Empire. (1910 newspaper ads for "Janos Hunyadi Natural Laxative Water" demonstrate the endurance of Hunyadi's fame.)

1531 A.D.   *Turkish armies continue to wreak havoc in Europe; bloody persecution of Protestants also rages in Europe; in the New World, Halley terrorizes the Peruvians, who read it as an omen of heavenly wrath. Before their fear has passed away, Pizarro arrives from Spain and destroys the mighty Inca Empire.*

1607 A.D.   *Holland is defeating Spain in her war for independence; Halley's appearance this year touches off a widespread panic that the end of the world is at hand.*

1759 A.D.   *War between England and France; General Wolfe captures Quebec; England is in the process of conquering India.*

1835 A.D.   *Carlist Rebellion in Spain; Indian War in Florida*

1910 A.D.   *?   ?   ?   ?   ?   ?*

Comet Dead Ahead!

In February, the usual kinds of background comet stories were interrupted with a new message, one more alarming than any that had come before. "Tail of Comet May Reach Us" blared the headline of a February 6 article, datelined Paris, France, and printed in newspapers everywhere. Like most of its competitors, the *Boston Herald* reported the startling news in a breathless fashion. "Astronomers Discuss Probable Effect of Contact with Our Atmosphere" read the subhead, and the text proceeded to discuss the possible outcome of such an encounter.

Dr. G. Bigourdan, of the French Academy of Science, is quoted: "The comet's atmosphere may touch that of the earth, but people will not be troubled thereby." Soothing words, to be sure. But the last paragraph of the article quotes another and equally eminent scientist, Dr. Henri Deslandres, director of the Astronomical Observatory at Meudon, France. "The spectroscope has proved the existence of *enormous quantities of cyanogenic gas in the tail of the comet.*" (Italics added.) After this astounding connection between the two facts that would soon be uppermost in the world's consciousness (poison gas in the comet's tail and our coming passage through that tail), Deslandres blandly continued, "I don't think, however, that the contact of the comet with the earth's atmosphere will be dangerous."

Here again is the kind of mixed message from the scientific community and the media that was bound to confuse the man in the street, and leave him vulnerable to the most extreme fears.

Surprisingly enough, the explosive implications of this item were not immediately apparent to news editors. But it didn't take long for them to get the message. One day later, the *Herald* ran the following item, giving it a little more space and placing it closer to the front page.

"*COMET'S TAIL IS POISON . . . Astronomers Believe It May Snuff Out Life on Earth*" (Italics added.) The reader is first informed that American scientists at Yerkes Observatory have proven that "the spectra of the comet . . . show very prominent cyanogen bands." In the following paragraph, the news is even worse:

May Poison Earth's Air

Cyanogen is perhaps the deadliest poison known, a grain of its potassium salt touched to the tongue being sufficient to cause instant death. In its uncombined state it is bluish gas, very similar in its behavior to chlorine, and extremely poisonous. [Italics added.]

In case any readers have still missed the point, they are treated to the considered words of Professor Camille Flammarion, "the distinguished French scientist, who is of the opinion that the cyanogen gas would impregnate the atmosphere and *probably snuff out all life on the planet.*" (Italics added.)

One day later, a third story appeared in the *Boston Herald*: "Harvard Scoffs at Fear of Comet!" The contents of the article, however, present another mixed message. On the down side, we learn that Professor Edwin Booth, head of the Chemistry Department at the University of California at Berkeley, is "forecasting wholesale fatality." Speaking for the up side is Professor W. H. Pickering, head of the Astronomy Department at Harvard: "Notwithstanding Profs. Booth and Flammarion to the contrary, the old earth has a long life ahead of it. In the last century it has passed through three comets' tails with such proficiency that it is safe to predict it will go unscathed through several more."

It was February, with three months to go before the comet could possibly do its damage. And three months for the enlightened population of the world to decide which group of experts it would believe. For the "unenlightened" of the world, the issue was basically a choice between superstitious fear and faith. But for the "enlightened," it was not even a simple case of science versus superstition—it was science versus science. After all, prominent and respected scientists were to be found on both sides of the issue. Would the world end in three months? Would everybody die? What was a person to believe?

The Drumbeat Quickens

In the eight month period between the recapture of Halley and Earth's encounter with its tail, Halley was the favorite topic of the international press. Every newspaper reader in the world was treated to at least one daily dose of Halley, and usually more. The comet was the subject of news articles, features, editorials, cartoons, letters to the editor, diagrams, and photographs.

As one example of the press response, consider the *New*

York Times. It began its sustained Halley coverage in August 1909, and ran at least one prominent piece per week for the rest of the year. In January 1910, the *Times* published over three Halley stories per week.

As the comet approached, the number of articles continued to increase. The *Times'* comet coverage peaked in the month of May, during which every edition of the paper, except two, contained one or more pieces about Halley, many of them on page one, and several in the headline spot reserved for the major story of the day. In recognition of the story's importance and the interest it held for the public, The *Times* also took the somewhat unusual step of assigning a journalist/amateur astronomer named Mary Proctor to the story. For five weeks, Proctor turned out a series of articles, some historical, some based on reports from around the world taken off the transatlantic cable, and some first person accounts of her own observations from atop the *New York Times* tower in Manhattan.

Despite its scope and depth, the *Times* comet coverage was not as large, and certainly not as sensationalistic, as the coverage of many other newspapers. But in all newspapers, the comet stories did not appear alone. They were surrounded by stories about everything else that was going on in the world. The public did not experience Halley's apparition as an isolated event. The general perception of the comet was colored by everything else that was going on in the world. Just as we will experience the 1985–1986 apparition within the context of our world (which will include space probes and TV pictures of the comet transmitted back to Earth), during the last apparition the comet was seen, felt, and experienced within the fabric of the world of 1910. To understand the public's response to Halley's last visit, we must make a visit to their world.

*New York street scene, circa 1910. Greeley Square looking north up
Broadway from 33rd Street. (The Bettman Archive, Inc.)*

Chapter 7

The World of 1910

At least two aspects of the world have remained fundamentally unchanged since Halley's last visit—the geography of Earth and the complexities of human nature. On other fronts, much has changed.

On the international scene, the British Empire was the strongest industrial and military power. Britannia indeed ruled the waves. But the German Empire, under the leadership of Kaiser Wilhelm, was making a strong bid for supremacy. In many fields, German technology was in the process of establishing itself as the standard of excellence.

Africa and Asia, for the most part, were divided up among the imperialist powers of Europe: Belgium had the Congo, France had North Africa and Southeast Asia (they called it French Indo-China), England had Canada, Australia, India, and more. Everyone was trying to get a foothold in China.

Virtually alone among the oriental peoples, Japan maintained its independence. What's more, with its victory over a bumbling, disorganized czarist Russia in the Russo-Japanese War of 1905, Japan had become an international military power, the only nonwhite power in the world. In the eyes of the Western World, Japan was seen as the Yellow Peril.

. . . .

Kings were still in fashion. With the exception of France, the United States, and the unstable nations of South and Central America, all of which were republics, the players in the international game of world politics were all monarchies. What's more, the kings, queens, czars, kaisers, and emperors were all related to each other and few if any belonged to the same national stock as the subjects they ruled.

The United States was still a young, untested giant, a raw but powerful rookie among the seasoned veterans. Everyone conceded to the United States her seemingly unlimited natural resources, and her growing industrial and agricultural strength, but we had not yet played a full hand in the world game.

In 1910, William Howard Taft was president. He was a large man, and at 6'3" and over 300 pounds, he was the largest ever to fill that office, but his shoes left little imprint on the trail of history. He is probably best remembered today as the chief executive who began the custom of throwing out the first ball on opening day of the major league baseball season.

His predecessor, Theodore Roosevelt, remains one of the most memorable of all presidents. He was "the great Trust Buster," the man whose motto was "walk softly and carry a big stick," He used the stick, too. At home, he aimed it at what he called "the malefactors of great wealth." His doctrine was the "Square Deal," and he tried to keep some fairness in the American capitalist system. Against opposition, he also led the nation to undertake one of the great engineering feats of all time, the digging of the Panama Canal.

Roosevelt left office in January 1909, after declining to run for another term, but in 1910 he remained unquestionably the most famous American in the world.

In 1910, radio was still ten years in the future, though the technology was already on hand, and broadcast television almost 40 years away, but the news was disseminated and public personalities were created, all in print. That year over 22,000 newspapers were published in the United States, about half of them in English and the other half in languages from all over the world.

Every major American city had several competing newspapers, each trying to take over the market and each representing a particular point of view. Serious news articles were long, informative, and frequently packed with a great deal of detail. Life was less hectic in general, and no other media competed for the readers' attention, so editors could safely make the assumption that their readers would and could read.

It was the tail end of the irresponsible, flamboyant period in the history of the American press known as the "Era of Yellow Journalism." In those days some newspapers would print anything to sell more copies. If a reporter couldn't get a good story, he was encouraged to make one up.

The news stories of 1910, like those of our own day, were a mixture of the serious and the absurd, from the development of a "safe" container for the transport of radioactive material (radium) to the invention of eyeglasses for chickens.

The Numbers of 1910

One way to get an idea about another time in history is to look at its numbers. Beyond being mere cold statistics, these numbers are full of human implications, for it took people to create them. At the very least, these figures will provide a framework for the story that follows. They may also begin to describe what life was like in the United States of 1910.

- The population was 90 million, which included the 8 million immigrants who had arrived since 1900.
- There were forty-six states in 1910, the newest being Oklahoma, which had gained statehood in 1907. In June 1910, the U.S. Congress passed an act to enable the people of Arizona and New Mexico "to form a constitution and a State government and to be admitted into the Union on an equal footing with the original States." They were duly admitted in 1912, bringing the count to forty-eight, where it remained until 1959 when Alaska and Hawaii were admitted.
- The world wheat harvest in 1910 was 3½ million bushels. The United States (737,000) and Russia (711,000) were the two major growers.

• • • • • • • • • • • • • • • • •

- Over 25 million tons each of steel and pig iron were produced, in each case over 40 percent of the world's total.
- Over 170 million barrels of crude oil were produced, well over 60 percent of the world's total output for the year. More and more of the crude was being processed into gasoline to fuel the growing fleet of automobiles, trucks, and mechanized farm vehicles.
- Animal power on the farm was rapidly being replaced by the gasoline engine. Production of tractors reached 4000 in 1910, by 1920 it reached 200,000.
- At the start of 1910, there were more than 200,000, "self-propelled vehicles" in the United States. By the end of the year, there were well over 400,000.
- In New York State, over 65,000 automobiles were registered, and over 20,000 chauffeur licenses were issued.
- From a negligible position in 1900, by 1910 the automobile industry had grown into the fifth largest industry in the United States.
- There were 1000 miles of concrete road in the United States, a sevenfold increase over the 140 that had existed in 1900.
- Dozens of independent manufacturers produced autos all over the country. Makes included Locomobile, Zust, Empire, Cadillac, Ford, Buick, Benz and Fiat (imports), Grout, Hudson, Isotta, Moon, Columbia, Simplex, and Pullman. A Hupmobile cost $750, a Reo runabout was $500, and a Sears, Roebuck Model L could be bought new for $370.
- Railroad magnate James J. Hill denounced the automobile as "an industrial menace" to drive and an "economic waste" to buy. With an eye to the possible implications of the automobile for his own industry, Hill's remarks may not have been entirely objective.
- Railroads were still the mainstay of American transportation in 1910. There were over 300,000 miles of operating railroad track, 60,000 locomotives, 36,000 passenger cars, 13,000 baggage and mail cars, and over 2 million freight cars.

. . . .

- In 1910, there were over 1½ million railroad employees in the United States.
- The transatlantic fleet of passenger liners, which sailed chiefly between New York and various European ports, numbered ninety-six ships.
- In September 1910, the *Mauretania* set a new Atlantic crossing record: Queenstown to New York in four days, ten hours, forty-one minutes.
- Professional baseball was by far the most popular sport. In terms of audience size, it was the biggest draw in the nation in the early 1900s, when the sixteen major league teams had a total yearly attendance of almost 7 million fans. This total does not include attendance figures for the thousands of minor league, semiprofessional, industrial, collegiate, and amateur teams around the country.
- Vaudeville, too, had been a mainstay of the entertainment scene in the early part of the decade, but a new entertainment was winning the public away. As early as 1907, the gross income of vaudeville and legitimate theater combined was exceeded by that of the new film industry. By 1910, movie attendance was reckoned at nearly 10 million per week.
- Big league baseball had been the source of many pop heroes of the day: almost everyone had heard of Christy Mathewson and Ty Cobb. But within a few short years, their fame was overshadowed by such worldwide screen personalities as Charlie Chaplin and Mary Pickford.

Entertainment and the Arts

Many Americans still made their own music, but by 1910 the hand-wound phonograph was invading the nation's living rooms. Hit songs of the year included "Heaven Will Protect the Working Girl," "Come, Josephine, in My Flying Machine," "Down by the Old Mill Stream," "Some of These Days," "Let Me Call You Sweetheart," and "Put Your Arms Around Me, Honey."

• • • • • • • • • • • • • • • • •

One of the year's hits was a sentimental Irish ballad. It had a coincidental link with one of the year's most sensational murders. The story began a few years earlier on St. Patrick's Day in Cleveland, Ohio, where Stanley Ketchel, the middleweight champion, was giving a boxing exhibition. He offered $100 to any man who could go three rounds with him.

A poor, young musician named Ernest Ball, in need of carfare to New York, stepped out of the crowd and into the ring. Ketchel, called "the Michigan Assassin," was known as the most savage of all prizefighters. For three painful rounds he battered and slashed the inexperienced musician, but Ball survived, collected his money, and went off to New York to find fame and fortune.

In 1910 Ketchel's ferocity caught up with him: the "Michigan Assassin" was shot to death on a farm in Missouri. Halfway across the country, in Tin Pan Alley, Ernest Ball composed his first big hit, "Mother Machree." And a short time later he wrote the Irish-American national anthem, "When Irish Eyes Are Smiling."

Victor Herbert's new light opera, Naughty Marietta, played for 136 performances at the New York Theater. Its hit songs included "Tramp! Tramp! Tramp!" and "Ah, Sweet Mystery of Life."

The Ziegfeld Follies featured the eighteen-year-old Fanny Brice, who was paid $75 a week.

In mid-May, the days of the comet, the New York theater scene was booming. At the Gaiety on 46th and Broadway, young John Barrymore was in the ninth month of The Fortune Hunter, his first starring role.

Headliner Eva Tanguay, for her appearance at the Bronx Theater, was billed as "the Comet of Broadway."

As it has frequently been, Paris was a source of much stage material that season. At Hammerstein's on 42nd Street, the audience was treated to the "Latest Parisian Pantomime Sensation . . . A Night in the Slums of Paris."

Meanwhile, out at the Brighton in Coney Island, where the daily matinee cost 25 cents, the featured act was "Paris by Night," starring the Frenchman Molasso, who was introducing his big hit, the French Apache dance, to America. Also on the bill, and mainstays of the Brighton's program, was the well-known comedy team of Bedini and Arthur.

Following Molasso's act, Bedini and Arthur staged a parody of the sexy French dance, featuring their assistant as the sultry but pliant woman of the Parisian underworld. The assistant, in his first featured role, wore a slinky dress. In the reigning vaudeville tradition, he was in blackface. But the makeup couldn't hide his bulging popeyes: it was the eighteen-year-old Eddie Cantor.

Out in Brooklyn, another young blackface comedian was also trying to make it in show business. This one would soon become the most famous entertainer in the country, but as a singer: Al Jolson.

The best-known black entertainer of the day was Bert Williams. To ease his acceptance by white audiences, he too followed the vaudeville custom of the day and wore blackface. But Williams was no mere hanger-on; he was a master of mime, song, storytelling, and was a great comic. American and European audiences delighted in his routines, and he earned top dollar. Like Fanny Brice, Williams also made his Ziegfeld Follies debut in 1910. But, despite his international success, Williams was still a vulnerable black man, subject to a range of racial slurs that were taken for granted. In a *New York Times* ad for the Colonial Theater, for example, Bert Williams is featured as "the biggest cinder in the smokery."

At the Alhambra, on Broadway, W. C. Fields was billed as "the Eccentric Juggler," but at another vaudeville house in Brooklyn, he was "the Tramp Juggler." Already a star on the international vaudeville and variety circuits, Fields was still several years away from his career as a humorist which would immortalize him in a series of unforgettable films.

A short while later, a startling and revolutionary musical score for a ballet called *The Firebird* by a twenty-eight-year-old Russian composer named Igor Stravinsky caused a riot on opening night at the Paris Opera House.

Among the symphonic works premiered that year were Gustav Mahler's "Eighth Symphony," Arnold Schoenberg's atonal "Three Piano Pieces," and "Mother Goose" *(Ma Mére L'Oye)* by Maurice Ravel.

Poster for the Johnson-Jeffries fight, July 4, 1910. (The Bettman Archive, Inc.)

Chapter 8

The Stars of 1910

The Great North Pole Dispute

On September 1, 1909, ten days before Halley was recaptured by Wolf in Germany, the world was electrified when Dr. Frederick Cook announced that he had won the race to be the first to reach the North Pole. But only six days later, Commander Richard Peary informed the world that *he* had been the first.

In the public imagination of 1910, the attainment of the pole was equivalent to the moon landing in our own day. Now, after years of failed attempts, not one but two men came forth to claim the great prize.

The world's press had a field day, led by the *New York Times* and the *New York Herald*. The *Times* had bought the newspaper rights to Peary's story, and the *Herald* had paid $20,000 for the rights to Cook's. Each paper supported its own man, and their stories were sold to other newspapers everywhere. If this wasn't enough, the controversy was fanned by the personalities of the two men: under close scrutiny, Cook's background seemed to have some shady passages; Peary was one of the nastiest and most egotistical of all explorers.

The world soon split into two camps. Everybody had a favorite, but the combined weight of the establishment forces on Peary's side (the *New York Times*, the National Geo-

• • • • • • • • • • • • • •

graphic Society, etc.) eventually tipped the scale in his favor and he was acclaimed the winner. Cook, burdened by a series of accusations, was labeled a liar and consigned to the slag heap of imposters and fakes. But the controversy is still alive. In recent years a book and a TV movie have made strong cases for Cook. Perhaps history will eventually reverse the earlier decision. But whatever the final judgment, during the 1910 season of Halley mania, Cook and Peary were two of the most famous names in the world. Their effect on comet fever, unmeasured and perhaps unobserved, was great. Their controversy, in which reputable men from all walks of life were to be found on both sides of the question, foreshadowed the shape of the more feverish controversy over the issue of Halley's cyanogen tail . . . Would it or would it not kill all life on Earth?

Houdini Flies Down Under

Halfway around the world, the great Harry Houdini was baffling Australian audiences. But his remarkable feats of escape were not the only thing on his mind. Flying was his obsession that year, and Houdini had sailed to Australia with his 33-foot-long French-built Voisin airplane. It was a typical craft of the time, built of wood and metal and sporting two 4-foot-long aluminum propellers. Its fuselage and wings were covered with cloth fabric, Overall, it looked like a giant, oddly shaped box kite.

Nobody had yet flown an airplane in Australia and Houdini, who was still learning to fly, set his mind on being the first. At five o'clock in the morning on March 15, 1910, Harry Houdini flew his Voisin around a makeshift airfield at Digger's Rest, outside of Melbourne. He flew at an estimated speed of 50 mph, and landed perfectly. A few days later, the hastily organized Aerial League of Australia presented him with a plaque in recognition of his deed.

The *Australian Punch* quickly editorialized on Houdini's flight. Whether "advertising or adventure was the reason for Harry Houdini's flight didn't matter." The effect, the forward-looking editorial continued, was to underline the importance

• • • • • • • • • • • • • • • • • • •

of the airplane, particularly in the worldwide armaments race: "The battles of the future will go to whomever is strongest in the air." Though fighters and bombers did not yet exist, at least two nations, Germany and the United States, were already producing antiaircraft guns to defend against them.

The airplane had been in existence for only seven years, since 1903 when the Wright brothers made their historic flight at Kitty Hawk, North Carolina. But for the first five years of the airplane's existence, it received little serious attention in the press and did not capture the public's imagination.

But by the time Halley arrived in 1910, the airplane had become one of the central metaphors in the public imagination, thought about by men as diverse as poets and generals.

Jack Johnson and the Great White Hope

The single biggest sporting event of 1910 was the heavyweight championship bout between champion Jack Johnson and ex-champion Jim Jeffries. The fight was scheduled for July, but the international ballyhoo was in full swing by May. The story surrounding the match generated so much heat that prefight coverage often appeared on the front pages of newspapers, side by side with Halley updates from astronomers and warnings of impending doom from cometophobes.

Jack Johnson knocked out Tommy Burns in 1908 and became the first black heavyweight champ in history. For the next seven years, until he lost the title to Jess Willard in a celebrated bout held in Havana, Cuba, Johnson was much in the national consciousness. For most of white society, Jack Johnson's success was felt as their humiliation. And out of this widespread sense of white disgrace grew the search for "the Great White Hope"—a fighter who could beat Jack Johnson and regain the pugilistic honor of the white race.

Each "white hope" was presented to the public as if he was St. George about to go forth against the dragon. But the dragon was formidable. As recently as 1984, a panel of boxing experts placed Jack Johnson in a very small group named "the greatest boxers of all time." So Johnson had little trouble with great white hopes. He handily defeated them all.

• • • • • • • • • • • • • • • •

After a couple of years of new white hopes who couldn't slay the dragon, an old white hope was pulled forward. With the promise of a large purse, and an additional fee for motion picture rights, ex-heavyweight champ Jim Jeffries was persuaded to come out of retirement, whip the upstart black man, and restore white honor.

As soon as the fight was announced at the end of 1909, the press ballyhoo began. From the very beginning, the fight was seen as a contest between the races. Virulent racism of the red-neck variety was certainly not unknown, but the commonest form was a kind of gentlemen's racism, typified by John L. Sullivan in his *New York Times* articles, the first of which appeared on the front page of the May 1 edition.

Known far and wide as "the Boston Strong Boy," John L. Sullivan had won and lost his title in the nineteenth century, but he was still an international celebrity in 1910. In May he was in England, where he was one of the biggest attractions on the entertainment circuit. Giving two shows a day, which consisted mainly of a monologue about his long career in the ring, Sullivan drew an audience of 40,000 people a week. And while in England, he was put under exclusive contract by the *New York Times* to regale its readers with his impressions of the "Great Fight."

Sullivan had no doubt that his opinions were shared by the great majority of his readers, so he, or his ghostwriter, was straightforward: "The fight between James J. Jeffries, representing the white race, and Jack Johnson, who now stands sponsor for the fighting qualities of the black man, is, to my mind, going to be one of the greatest battles, if not the very greatest, in all ring history." With the sociological aspects of the fight uppermost in his mind, Sullivan continued, "When I said that the contest is really between representatives of two races rather than between two individuals regardless of the race question, I felt certain that many would undoubtedly share my opinion."

And in case his readers have still missed the point: "I myself feel very sorry the match was ever made. I am not biased, but I do believe that the negroes should fight in a class by themselves. Many times during my career I was urged by

outsiders to throw reason to the winds and fight a black man. But I always refused."

As for the international interest in the fight, Sullivan found that in England's "leading hotels it is the one absorbing topic of conversation among men in all walks of life. I believe," he added, "there will be five million dollars wagered in England on the success of Jeffries."

With a promise to share his knowledge and views with his readers in following articles, the writer adds, "of course, we shall all like to see the white man win, but wishes can never fill a sack. Should the negro win there is not a white man in the world who could throw up his hat and dance with glee . . . . I bear no malice toward any man living and I have fear of none. I know that in this case the public wants the cold, unvarnished facts. I am going to travel 7,000 miles to place those facts before the readers of the *New York Times*. Believe me to be, ever yours truly, John L. Sullivan."

The big fight was beset with problems, all of which served to increase public interest. One issue was location; after several communities turned them away, the promoters finally reached an agreement with Reno, Nevada, which was known as a wide open town. Another issue was the selection of the referee. It was commonly assumed that no man could be found who would be racially impartial. Many voices were raised in support of Teddy Roosevelt: he was a well-known sportsman with a particular interest in boxing, and his integrity was unassailable. It is not clear if the offer was ever made.

One internationally famous man was invited to referee the fight, Sir Arthur Conan Doyle, creator of Sherlock Holmes. Doyle was also a great fight fan and an amateur boxer himself, taking weekly lessons from a professional. When he received the offer to referee the fight, he was elated: "By George," Doyle said, "this is the most sporting proposition I ever had!" When asked if he'd go, Doyle replied, "Go? Of course I'll go! This is a real honor!" But after thinking it over, Conan Doyle had to decline.

The Johnson-Jeffries fight took place in Reno on July 4, 1910. The city's population of 15,000 doubled by fight time, with an influx of sporting men from all over the United States and from as far away as Canada, England, France, South Africa, and Australia. Tex Rickard, the fabled boxing promoter, was himself the referee: the managers of the two fighters could agree on no other man who was willing to take on the job.

The fight was originally scheduled for forty-five rounds, but it was stopped in the fifteenth after Jeffries absorbed a terrific beating. Both fighters were well compensated, each coming away with over $150,000 tax free.

Surrounded by more media hype than any sporting event had ever had, the fight came and went and a black man was still heavyweight champion of the world. John L. Sullivan's prediction that no white man would "dance with glee" at a Johnson victory proved to be an understatement. As news of Johnson's victory spread, there were antiblack riots in numerous American cities, including Boston, New York, Cincinnati, Houston, and Norfolk, Virginia. In Uvalda, Georgia, three blacks were murdered.

The postfight race riots were a reminder of an enduring American social problem, but the year of 1910 also witnessed a major step forward—the establishment of the National Association for the Advancement of Colored People (NAACP). Forty-three years later, in 1953, the NAACP brought the lawsuit which resulted in the Supreme Court's School Desegregation decision, surely one of the most far-reaching court decisions in American history.

Buffalo Bill Says Good-bye

Four days before the fateful encounter with Halley's cyanogen tail, another "hairy star" flashed into New York City. This one's hair was silvery white and fell to his shoulders. Instead of riding the gravitational pull of the sun, this one rode a big white horse. He was Buffalo Bill Cody, and he had just brought his Wild West Show into Madison Square Garden to begin his national "Farewell Tour." After thirty-four years in show business, the "Chief of the Scouts" was getting ready to call it a career.

Dressed in a stylized version of his scout's costume of the

• • • • • • • • • • • • • •

plains, Buffalo Bill galloped into the arena on his white horse, his silvery hair streaming over his shoulders. He was greeted with five solid minutes of wild cheering from the 7,000 men, women and children who had come to see him. When the crowd quieted down, Buffalo Bill made his well-known introduction: "Ladies and Gentlemen, allow me to introduce to you the Congress of the World's Rough Riders!" Another round of mighty cheers filled the Garden.

The rest of the cast, over a hundred strong and most of them on horseback, paraded around the arena. When they left and he was once again alone in the spotlight, Buffalo Bill gave his farewell speech: "I have always looked forward to coming to New York every year," he proclaimed, "and seeing the smiling faces of the women and children welcoming me. It has been a great gratification to me to see how the show has been appreciated in the East by the rising generations, and the interest that has been taken by you all in the buffaloes and the Indians which have now disappeared from the plains to make way for the homesteads of the thousands who have gone West. Once more I thank you all for your kindness and say good-bye!"

With a flourish of his cowboy hat in all directions, Buffalo Bill slowly backed his white horse toward the exit, while all the people stood and cheered, some with tears in their eyes. Then the Wild West Show began.

The scenes were familiar, but no less rousing. They included the "Attack on the Deadwood Stagecoach," the "Battle of Summit Springs," and the "Immigrant Wagon Train in the Desert," which featured Buffalo Bill himself. The audience cheered every stunt, every bit of horsemanship, every act of markmanship, and the death of every Indian in the staged battle scenes.

In the spring of 1910, Buffalo Bill Cody was an American institution. He was popularly regarded as the last link with the old frontier of the plains, and to many people, his stage-shows were the real thing. He was surrounded with the aura of adventure: Indian scout, frontiersman, buffalo hunter. Newspapers found him to be an unending source of good copy. He was the hero of approximately 300 "dime novels," the trashy paperbacks of their day, and he was listed as the "author" of many of them. A dozen years earlier, when Theodore Roosevelt needed a name for the troop of horsemen he led up San Juan Hill in the charge that would lead all the way to the White House, he looked toward Buffalo Bill's Wild West Show and called them the "Rough Riders." What's more, Cody's European tours had given him an international following.

In a long and sympathetic interview published the next day in the *New York Times*, Buffalo Bill spoke of his retirement. He said he planned to oversee his investments, which he claimed included "one hundred and four gold and silver claims," and thousands of acres of land.

Perhaps it's odd to think of Buffalo Bill as a land speculator and developer, but here is what he said: "Every cent that I have made in the show business I have invested in the West in developing the arid plains that are now fine home lands peopled with happy American families. Once I spent $700,000 in digging an irrigation canal before I got a cent returned. Some of the land I bought then for $2 an acre I sold last week for $480 an acre, but then I hung on to it, don't you see. As the buffalo died out and the Indians were put on the reservations, I put my money into the arid land to irrigate it and convert it into home land."

There is no telling how much was fact and how much fancy in Buffalo Bill's words. The truth is, he was a fictional character posing as a real person. The character known as Buffalo Bill had been created, mostly out of whole cloth, by publicists and hacks. The tragedy lay in Cody's inability to distinguish between the fiction and the fact. His career as a scout and buffalo hunter had ended when he was still in his twenties, and he spent most of his life as a showman. According to one social historian, so much myth was infused into history that "no completely reliable biography of Cody has been or ever will be written."

Reality began to catch up with Cody in 1910. After a long marriage, he sued his wife for divorce on the grounds she was trying to poison him with "dragon's blood." For five years he wrestled with the problem, first filing for divorce, then with-

. . . .

drawing the suit. Finally, around the time he began his farewell tour, he and his wife were reconciled. She said, "I love Will. He is the kindest, most generous-hearted man I ever knew." As for the "dragon's blood," it turned out to be a love potion she'd used when everything else had failed.

Buffalo Bill's farewell tour really never ended. For the last years of his life, mounting debts forced him to continue with the Wild West Show and other promotions. Toward the end, he applied to the Federal Government for the $10 monthly stipend given to recipients of the Congressional Medal of Honor.

Caruso Thrills Paris

On the night of May 19, the New York Metropolitan Opera made its European debut in Paris. It was a rehearsal performance of *Aida* before an audience which included the Premier of France and many other notables. The conductor, who received a remarkable ovation, was Arturo Toscanini. The role of Rhadames was sung by the most famous opera star of all, Enrico Caruso.

Earlier in the month Caruso had sailed from New York aboard the *Kaiser Wilhelm II* with a group of divas that included Geraldine Farrar and the Mademoiselles Fremstad, Gadski, and Destinn. The ship was nicknamed the "High-C Express."

Caruso, at the peak of his earning capacity, received no less than $2500 per performance. He carried with him his earnings for the American opera season, over $200,000, $50,000 of which was for phonograph record royalties.

While at sea, he received some welcome news via a recent invention, shipboard wireless. A young Italian from Brooklyn who had tried to extort $15,000 from Caruso was sentenced to three to seven years in Sing Sing Prison. Upon sentencing, the judge said, "You and your associates belong to a notorious Mafia society, which has no regard for life or property, and would as soon destroy the world's greatest singer and the pride of the Italian race as any other unfortunate victim."

Teddy Roosevelt leaves Africa and heads for his triumphal tour of Europe. (From the Theodore Roosevelt Collection, Harvard College Library.)

Chapter 9

Household Names Around the World

Two larger-than-life-size men occupied the center of the world's stage during the days of Halley in the spring of 1910. One was an Englishman, the other an American. They were very different kinds of men, but through coincidence, they had several things in common. Each had come to power in 1901 when his very popular predecessor died. Their rise to power was greeted with dismay by the ruling classes of their respective nations. Yet both achieved a worldwide popularity of immense proportions. And when the Englishman died in 1910, the American represented his nation at the funeral. The Englishman was King Edward VII. The American was Theodore Roosevelt.

Prince Edward the Long Heir

When Queen Victoria died at the age of eighty-two in 1901, the book was closed on the longest reign in English history. For sixty-four years, Victoria had been the symbol of England, revered and loved by her adoring subjects. And for sixty of those years, Edward, Prince of Wales, her son and heir, had waited in the wings.

· · · · · · · · · · · · · · ·

Under her long reign, the British Empire expanded into the largest land empire the world had known since Roman times. British world power was at its peak. Since Victoria's days, Empire and power have both shrunk.

Another diminishing legacy left by the Queen was the world view and lifestyle which bears her name: Victorianism.

Like all world views, Victorianism meant different things in different contexts. In politics it meant colonialism and imperialism (Victoria became Empress of India in 1876). In economics, it translated into industrialism and international trade. In architecture, it was expressed in homes with ornate exteriors and gloomy insides. Somehow or other, the interiors of typical Victorian homes seemed to symbolize the high, stuffy seriousness of life.

In the most important area, the conduct of one's life, Victorianism imposed the strict values of propriety, decorum, and restraint. Outward appearance was all important. What a person did in private, by way of vice or eccentricity, was allowable or at least conveniently ignored, so long as it remained private. Public behavior, however, had to conform strictly to the prescribed etiquette.

Victorian etiquette was elaborate and strict. There were rules for everything. The most elaborate rules of all were applied to sexuality. Not only was sexual behavior to be suppressed, its very mention was forbidden. Even words which might allude to sexuality, however remote, were not approved. For example, it was considered indelicate at the Victorian dinner table to offer the "leg" of a chicken to a lady. This led to the tradition, which is still alive, of offering her the breast, though the Victorians referred to it as "the bosom" of the chicken.

To the Victorian libido, "legs" were highly charged parts of the anatomy. In some circles, even the legs of pianos had to be covered with crinolines, lest the sight of them give rise to lascivious thoughts. The act of concealing a piano leg is to sexualize it, and though this seems a farfetched thing to do, it is fairly indicative of the Victorian obsession with the public denial of sex.

. . . .

For the last third of the nineteenth century, however, there was already on the official Victorian scene a man of consequence who seemed to behave as if sex was a natural part of life. Furthermore, he behaved as if sham and pretense were not necessary. That man was Edward, Prince of Wales.

As soon as he was able, Edward began his lifelong career as a womanizer on a royal scale. To be sure, the rest of his activities were also carried out in a larger-than-life fashion, but it was his womanizing that drew the most attention. His many love affairs and one-night conquests seemed to be common knowledge, much to the discomfort of his mother, who was deeply offended. As the royal spokesman put it, "The Queen was not amused."

Though he was her eldest son, and heir to her throne, Victoria had little respect for Edward's abilities and did not trust him. For most of his adult life, he was purposely kept out of all official state business. He did participate in the public pomp and circumstance of the royal household, but he received no state assignments from his mother. Most of his time and effort went into his pursuit of pleasure, and sensual pleasure at that.

Edward loved horse racing and yachting. Hunting, in the royal version of that sport, was another of his favorites. At one time or another, he shot elk in Sweden, bear in Rumania, tigers in India, elephants in Ceylon, and crocodiles on the Nile in Egypt. On a one-month hunt at a game preserve in Hungary, he and his entourage shot 37,654 head of game, which included 22,996 partridges.

Such hunting was done for "the sport" of it, though some of the game was sure to arrive on Edward's plate. His appetite for good food was prodigious, and he probably ate more than even the fabled King Henry VIII.

Edward's favorite breakfast included bacon, poached eggs, haddock, chicken, and woodcock. A late morning snack, to hold him over until lunch, would typically include lobster salad or cold chicken. His lunches ran to ten courses or more. Afternoon high tea included cakes, rolls, scones, crumpets, and tarts. Then dinner, usually a twelve-course meal. Finally,

. . . .

he ended the day's eating with a bedtime snack brought to his apartment, often a plate of sandwiches, a game bird, or a cold cutlet.

Edward's feverish pursuit of pleasure set the pattern for a growing section of upper-class English society. By the 1870s, while still Prince of Wales, he was the acknowledged leader of the international "smart set."

Edward's girth was not offensive to women, who seemed to find him irresistible, but whether they yielded to his prowess or his position is impossible to tell. While at home in England, he was relatively discreet, and many of his trysts were consummated at the homes of his *aides de camp*, left empty for his convenience. But while abroad, he was more carefree, frequently being seen in public with women on his arm or at his table at Maxim's in Paris, and none of whom were his wife, Princess and later Queen Alexandra. The English, as we've seen, had a slightly ambivalent attitude toward the romantic exploits of Edward. Some were offended, but others took a vicarious pride and pleasure in his uninhibited exploits. The French, always more subtle in their approach to these matters, displayed great finesse.

When Edward finally became king, some of his subjects were still apprehensive. Their fears were not diminished by an episode which occurred during Edward's coronation. By long tradition, the audience assembled in Westminster Abbey to witness the coronation of an English monarch is limited to the nobility. But in the King's Box, a section of the highest distinction, sat a small group of his past and present women, including Lady Jennie Churchill (Winston's mother) and Mrs. Alice Keppel, Edward's longtime favorite.

As king, Edward obviously had the power to invite anyone he wanted to his coronation. But some commentators remarked on his insensitivity to public opinion, and others felt he was flaunting his affairs. There was another point of view. Edward, as compared to the typical Victorian, was not a hypocrite. He was what he was, and never tried to pretend otherwise.

Despite his reputation, Edward set most of his subjects at

ease as soon as he ascended the throne. He took his royal
duties seriously, quickly demonstrating a talent in foreign
affairs.

His attitude toward the European continent was based on
his admiration for the French and his growing disenchant-
ment with Germany, which some said stemmed from an insult he'd received from his nephew, Kaiser Wilhelm. There is
plausibility in this personal interpretation of historical
events. Edward was always known as a man who never forgot
or forgave an insult. Eventually he led the push toward an
alliance between England and France—the "Entente Cor-
diale" which made those two nations allies in the event of an
attack on either by Germany. This alliance held through two
world wars, and is still in effect today. At the time, it was
optimistically seen as a great step toward preventing war in
the immediate future, and Edward won praise as a peace-
keeper.

With Edward's quick demonstration of his truly royal bear-
ing and his statesmanship, the critics of his private life grew
less strident, though he never altered his pleasure seeking
course, nor did he go far out of his way to pretend otherwise.
His long-term liaison with Mrs. Alice Keppel was well
known. She made frequent public appearances at the palace,
and the public nature of her relationship with the king was
reflected in the words of a popular song:

>"There is peace within the palace
>At a little word from Alice.
>SEND FOR MRS. KEPPEL!"

Edwardian society, particularly high society as represented
by the "smart set," shaped itself to satisfy his tastes. Com-
menting on Edward's reign, one historian summed up
thusly: "More money was spent on clothes, more food was
consumed, more horses were raced, more infidelities were
committed, more birds were shot, more yachts were commissioned, more late hours were kept, than ever before.

. . . .

It was, in short, the most ostentatious and extravagant decade that England had known."

Edward fulfilled his duties and won the respect and affection of his subjects. His power as a role model increased and solidified. His straightforwardness helped to prepare the way for the sexual revolution that swept through the Western world in the 1920s and has continued to the present time.

T. R.: The Rough Rider President

On September 6, 1901, President William McKinley made an appearance at the Pan-American Exposition at Buffalo, New York. The exposition had been drawing large crowds: some came to gawk at the pseudo-Spanish buildings which were supposed to symbolize our interest in Latin America. Others came to gasp at the beauty of the Electric Tower, a 400-foot-high structure illuminated with 35,000 light bulbs. It was the first-ever attempt at an outdoor electric display and it was a grand success.

The exposition was a hit, but McKinley's visit stirred additional excitement. He was an extremely popular president, and he was still riding the crest of our victory in the Spanish-American War three years earlier, a victory which had given us possession of Puerto Rico and the Philippine Islands with their 7 million inhabitants. In the expansionist, imperialist mood of the times, he was generally perceived to be a just and a good man.

McKinley was also a masterful politician. He was one of the first great handshakers, and had been clocked at fifty shakes per minute. So when he arrived in the Temple of Music to make a speech, he couldn't resist the opportunity to shake a few more hands. While an organist named Gomph played Bach, a long receiving line filed by. When one of the men in line, a self-styled but unaffiliated anarchist named Leon Czolgosz, approached the president, he extended a bandaged hand, but not for a shake. The bandage concealed a nickel-plated .32 caliber revolver, and without a word he shot McKinley twice in the stomach at point-blank range.

Mad confusion and rage followed. Czolgosz was mauled

. . . .

and beaten, and had to be smuggled out of the fair grounds to save him from an instantly formed lynch mob.

McKinley was rushed to the small emergency hospital on the fair grounds. Within two hours, he was put under ether. The operation was performed by a gynecologist, who was chosen because he was the only doctor in attendance known to the president of the exposition. He had never operated on a bullet wound before. While two well-known and highly experienced surgeons stood by, the gynecologist opened the president's stomach. He did not find the bullet, did not discover the serious damage to internal organs, did not drain the wound, and sewed him up. Eight days later, President William McKinley needlessly died of gangrene poisoning, and Vice-President Theodore Roosevelt was sworn in to take his place, becoming the youngest American president in history.

Theodore Roosevelt had to be summoned off a mountain in Vermont, where he'd been hiking, to rush to Buffalo to be sworn in. The country was not at all surprised: he was well-known as a man of direct action and boundless energy. Indeed, the night he took the oath of office, he was already a national hero, known to everyone as TR or Teddy.

In his presidential years, TR never slackened his energetic pace. He continued to take his rigorous "point to point" walks, on which every obstacle had to be scaled, climbed over, or swum across. He took boxing lessons in the White House until he accidentally lost the sight of one eye, so he switched to jujitsu. His interest in outdoor life led him to champion the birth of the National Park System. Seeing no contradiction, he also continued to be an avid hunter.

Soon after taking office, TR went on a bear hunt. For millions of children, it was the most significant bear hunt ever, and without knowing it, they cuddle up with its outcome every night.

On that hunt, Roosevelt was indeed confronted by a bear, but it was only a small, shivering cub. Naturally, he refused to shoot it. This gesture inspired Clifford Berryman of the *Washington Star*, who drew a cartoon of the scene—Roosevelt, rifle in hand, turning his back on the little bear.

The widely syndicated cartoon, in its turn, inspired a Rus-

• • • • • • • • • • • • • • • •

sian immigrant named Morris Michtom, who owned a candy store in Brooklyn. As quickly as possible, he took some brown plush, sewed it into the shape of a small bear with moveable limbs, stuffed it, and sewed on some buttons for eyes. Then he plunked it down among the penny candies in his store window and offered it for sale as "Teddy's Bear."

It was a big hit. Michtom later founded the Ideal Toy Company. And the Teddy Bear, as it's now universally known, has become immortal. It is safe to say that Teddy Roosevelt, or at least some unknown aspect of his aura, has received more cuddles and been taken into more beds than any president in history. He is likely to retain that distinction in the forseeable future.

TR's national fame was made with his cavalry charge up San Juan Hill, in Cuba, at the head of his Rough Riders. It was one of the few dramatic moments in the otherwise anticlimactic Spanish-American War of 1898. The other big moment was provided by Admiral John Dewey when he defeated the pitiful Spanish fleet at Manila and took possession of the Philippine Islands. Dewey and TR were the public heroes of that war, along with newspaper publisher William Randolph Hearst, who was the chief fanner of the war flame.

After the war, TR was elected governor of New York. The public adored him, but the men in control of the state's Republican Party did not trust him; they thought his ideas were too radical and not good for business. So they got him nominated to run for vice-president on the McKinley ticket of 1900. Marc Hanna, the major power broker in the national Republican Party, also mistrusted Roosevelt. Right after the nomination he cried, "Don't any of you realize that there's only one life between that madman and the presidency?"

After McKinley's death and his own fall from power, Hanna still found time in the midst of his grief to chastise his Republican Party colleagues: "Now look, that damned cowboy is president of the United States!"

As TR felt his way into the presidency and began to act as his own man, he showed that Hanna and his cohorts were right to worry about the "damned cowboy." In place of the

. . . .

Republican conservatism of the day, which McKinley had more or less typified, TR joined forces with the Progressives. Instead of defending the status quo, he instigated change and reform.

The old-guard establishment leaders, who thought at first that TR would mean business as usual despite his rhetoric, were soon disabused of that hope. In one of his early moves against the combines, monopolies, trusts, and giant corporations which were conspiring to inhibit competition in the supposedly free market place, TR invoked the Sherman Anti-Trust Act. Though his is the name that is most closely associated with it, Roosevelt had no part in the enactment of this law. It had been passed twelve years earlier, in 1890, after a great public outcry against the growing power of the trusts, but it had rarely been used. In fact, the few times it had been invoked, it was used against labor unions, not trusts.

The combine TR chose to battle was the Northern Securities Company, which had a capitalization of $400 million. The size of its funds was not a problem, the issue was constraint of trade. The Northern Securities Company came into being as the result of a draw in one of the great capitalist battles of the early part of the century. E. H. Harriman, the brokerage firm of Kuhn, Loeb, and Company, and John D. Rockefeller were on one team; the other had James J. Hill and the House of J. P. Morgan. At issue was control of all the railroads west of Chicago, and perhaps of all the railroads in the country.

After an intense financial struggle, which caused a brief panic on Wall Street, neither side could claim victory, so they merged forces and combined all the different individual railroads which had been at issue into one giant corporation, Northern Securities. Now that they had reached a peaceful settlement, these powerful men settled back. It never occurred to any of them that the federal government would object to their marriage of convenience.

When J. P. Morgan heard the news by way of a phone call from a reporter, he was shocked. That the government was planning to invoke the anti-trust law surprised him, and some of his associates refused to believe it was true just because it

• • • • • • • • • • • • • • • • • •

was so extraordinary. What shocked Morgan was that the president was acting in public rather than private. So he went to see Roosevelt in the White House and said, "If we have done anything wrong, send your man to my man and they can fix it up." TR, of course, didn't want to fix things up, he wanted to stop them. That is what he told J. P. Morgan, and the great financier left in a rage.

The Northern Securities case was settled in the government's favor two years later, but the eventual outcome was less important than the lasting impression TR made on the hearts of his countrymen. From here on, he was known as the "Trust Buster," the leader who stood up for the rights of the "small fellows."

The financiers, the men TR now called "malefactors of great wealth," had a different point of view. The president was not following the unwritten rules which had traditionally set the style for the relationship between government and big business. They were offended, angry, and in some cases, frightened. According to one newspaper in Detroit, "Wall Street is paralyzed at the thought that the President of the United States would sink so low as to try to enforce the law." That the financiers neither forgot nor forgave was demonstrated when they effectively blocked TR's bid for the Republican presidential nomination in 1912, thus paving the way for the victory of Democrat Woodrow Wilson.

Theodore Roosevelt's grip on the American imagination has been matched by only one other president in his century, his cousin Franklin Delano Roosevelt. To be sure, the Camelot days of John F. Kennedy were in the same league, and because of the electronic media perhaps even more intense. But as the focus of the national imagination, Kennedy was a relatively short-lived phenomenon. FDR was the focus for his twelve and one-half years in office, and TR held center stage, in or out of office, from the moment he led his Rough Riders up San Juan Hill in 1898 until World War I.

In our age of instant celebrities, who fade as quickly as they appear, it is difficult to understand the intensity and duration of TR's fame. Why was he so popular for so long? Part of the

reason is the quantum increase in the pace of life between his time and ours. Then, the rate of change was slower in most things, including taste, and heroes simply lasted longer. Another had to do with the perfect matchup between TR's image and the needs of the American imagination.

The first decade of this century was a time of clashing currents. The old order was being challenged by the new. In Europe the old was represented by royalty and capitalists, in America by capitalists alone. Old ways of doing things were also changing.

The world of 1910 was already well into the Age of Technology, itself another phase of the Industrial Revolution, which had drastically changed the way the mass of people lived their lives. Looking backward from 1910, within the span of a single human lifetime, the new "white magic" of technology had already pulled out of the hat of progress such previously unimaginable gifts as electric lights, automobiles, airplanes, the telephone, the radio, plastic, synthetic fiber, radium, motion pictures, the phonograph, the chain saw, the gyroscope, the Geiger counter, antiaircraft guns, and instant coffee.

It is safe to say that in 1910, technology and the idea of progress were two of the items on any list of things average Americans believed in. They also believed in more mundane things, like the right to earn a decent living, the right to feed and clothe their families, the right to make a better life for their children.

Industrialization and technology had created more wealth than had ever existed before in history, but this wealth was controlled by a small minority. For example, in 1910, fewer than 200 corporations (not counting banks) possessed over 40 percent of all corporate assets. The small business man, the little entrepreneur trying to make it on his own, was still held up as a model of the American virtue of independence, but the fact is, small, independent businesses were being forced out by the large corporations and trusts.

As for the workers, the ones called "wage slaves" by the "radicals," their lot was even worse. The average American mill worker earned about $10 a week, but wealthy society matrons could afford to lavish tens of thousands of dollars on a single night's entertainment.

TR was in tune with his times. When the country was for war, Teddy Roosevelt was a war hero, and the nation loved him for it. When the trusts were crushing the little man, TR attacked the trusts, and the nation loved him again. Throughout his entire public career, he was seen as the rugged individualist who cared, the maverick who was for the common man.

Part Three

In the Glow of the Comet's Tail

Making fun of Herr Doktor Sigmund Freud. (The Bettman Archive, Inc.)

Chapter 10

The Power of Perihelion

For practical purposes, astronomers mark the beginning and end of a comet's orbit at the moment of perihelion, when it makes its closest approach to the sun. At its furthest point, Halley is some 3,310,000,000 miles from the sun. (That was its location in 1948 when it turned and headed back toward us.) During perihelion, it is some 50,000,000 miles from the sun, a relatively close passage, which astronomers can measure with a high degree of accuracy.

For astrologers, perihelion is a powerful time in which the forces of the heavens are active and full of significance. In 1910, Halley's comet reached perihelion on April 20. On that day, an interesting but obscure event took place in Vienna.

Vienna was still an imperial city, the capital of the Austro-Hungarian Empire, which stretched from central to south-eastern Europe and ruled over some 50 million people. The city itself, with over 2 million inhabitants, was the sixth largest in the world.

On the surface, Vienna was a contented city. The middle class success and self-satisfaction of the Viennese—some called it smugness—was expressed by the monumental buildings erected in the Ringstrasse in the preceding quarter cen-

· · · · · · · · · · · · · · · · ·

tury. Afternoons were for long gatherings at the famous coffee houses, where the stout citizens sipped coffee topped with whipped cream and nibbled cream-filled Viennese pastries. In the evenings, glittering crowds gathered at the opera or the theater, and the cafes came alive with animated conversation, much of which, during the spring of 1910, was about a revolutionary set of ideas that had sprung up like a sour note in the home of the Viennese waltz.

The intellectual life of Vienna was in ferment over the startling ideas of Herr Doktor Sigmund Freud, founder and leader of the new psychoanalytic movement. From his residence at 19 Berggasse, Freud directed the early struggles of his growing movement. Many members of the Viennese medical establishment thought he was a crackpot. Outrage was also expressed by members of the American medical establishment.

In May, at the annual meeting of the American Neurological Association (ANA) in Washington, D.C., one Dr. Joseph Collins rose to denounce the ideas of Freud and his followers, labeling them "pornographic stories about pure virgins." Collins, whom Freud's biographer, Ernest Jones, tells us was notorious for his dirty jokes, went on to say "It is time the ANA took a stand against Transcendentalism and Supernaturalism and definitely crush out Christian Science, Freudism, and all that bosh, rot, and nonsense."

But despite the increasing hostility, Freud's ideas found fertile soil in the United States. Immediately following the ANA meeting, a dissident group of medical men banded together to form the American Psychopathological Association, the forerunner of the American Psychoanalytic Association, with Freud himself as an honorary member. His ideas had found a permanent forum in America.

Sigmund Freud, whose ideas would change our view of human nature, was already on the world stage in the time of the comet in 1910. Another resident of Vienna, who would also leave his mark on history, was still unknown to the world-at-large.

On the other side of the city, far from the Ringstrasse and

. . . .

the haunts of cafe society, were the worker's quarters in the grim industrial area along the Danube. At 27 Meldemann-strasse, a young drifter shared a flop-house dormitory with several other down-and-outers. He was a strangely intense young man. His long black hair hung over his greasy coat collar. What could be seen of his face was very pale; the rest was covered with a black beard. His eyes were his most telling feature, large and oddly hypnotic. His name was Adolf Hitler. Years later, people who knew him at this time of his life always remembered his eyes—how they stared and glowered, particularly when he was excited.

Hitler had come to Vienna from the city of Linz to be an art student at the Vienna Academy of Fine Arts, but his application had been twice rejected. He was embittered, though he let his acquaintances back in Linz think he was leading the carefree life of an art student. In reality, he was trying to eke out a living by making stiff miniature water color paintings. They were copies of other artists' "views of Vienna," and they were peddled in the beer halls of the city by Hitler's partner, another drifter trying to survive.

By the spring of 1910, Hitler had grown tired of making the little paintings. He began to spend more time reading. In the newspapers he focused on the political news. At the public library he delved into an assortment of subjects and grew interested in occultism. He developed an intense interest in astrology which remained with him until the end of his life.

Two things happened to Hitler simultaneously that spring. One public and one private. The public event was his debut as a political rabble-rouser, which began on a small scale in beer halls and in his flop-house dorm. The private event, shaped by Halley's comet to an extent we will never know, was his twenty-first birthday. Hitler, in the midst of his first enthusiasm with astrology, turned twenty-one on the most highly charged day of all, April 20, the day Halley reached perihelion. There is no way to gauge how deeply affected Hitler was by the astrological coincidence, and we can only be left to wonder if Halley's comet played a role in the most destructive career of the century, perhaps of all time.

• • • • • • • • • • • • • • • •

In and Out with Halley's Comet

The very next day, April 21, Mark Twain, the most beloved American humorist, died at his home in upstate New York. Cometophobes found great significance in his death, but not as much significance as Twain himself.

In its 1835 apparition, Halley had reached perihelion on November 14. Two weeks later, as Halley began its majestic outward passage to the edge of the solar system, Samuel Clemens was born in Florida, Missouri. The boy who was to become Mark Twain had a lifelong fascination with Halley. He believed it was his talisman.

By the end of 1909, Twain was an internationally loved author and lecturer, but he was a bitter and tired man. Two of his four children were already dead, as was his beloved wife, Livy. His finances were shaky and he was ill. At Christmastime, his daughter Jean died suddenly. He never recovered from this blow. His health failed rapidly; he lost his desire to live: "I have never greatly envied anyone but the dead. I always envy the dead."

A short time later, in a letter to a friend, he imagined God to be talking about Mark Twain and Halley's Comet, his old talisman which was once again crossing the skies of Earth: " 'Here are those unaccountable freaks. They came in together, they must go out together.' Oh, how I am looking forward to that." On April 21, 1910, the day after Halley reached perihelion, Mark Twain died at the age of seventy-five.

An outpouring of articles and editorials followed Twain's death. Most were laudatory, but one magazine wondered "Whether *they* will have permanence in popular esteem which is accorded to the classics may well be doubted." The "they" referred to were Tom Sawyer and Huckleberry Finn.

Many of Twain's political quips were reprinted, particularly those aimed at Congress. Perhaps this was a reflection of the national mood. Newspaper readers all over the country were delighted to read, "Congressmen are the only distinctly native American criminal class." Or "Fleas can be taught almost anything that a Congressman knows."

•   •   •   •   •   •   •   •   •   •   •   •   •   •   •   •

Mark Twain was gone, but his mighty talisman lingered on. It was just beginning its outward passage.

The Canals of Mars

Around the time of perihelion, one of the best-known astronomers in the world was astounding audiences with his strange tales of the planet Mars. Professor Percival Lowell of the Harvard University Astronomy Department was on the lecture circuit spreading his theory about the canals of Mars. He filled auditoriums in cities such as London, Berlin, and Paris. With Halley overhead, interest in all matters astronomical had soared, and Lowell's story of the Martian canals was particularly relevant.

Lowell based his theory on evidence that was much disputed: a few fuzzy telescopic photographs of the planet Mars. He claimed that a faint spiderweb of lines were too geometrical to be natural phenomena. Most astronomers dismissed the whole idea as not worthy of serious discussion. But Lowell, after all, was at Harvard, and in his lectures and monographs, he firmly stated that the canals did in fact exist. The implication of the canals, if they truly existed, was overwhelming. If they were there, somebody had to be there to build them.

The suggestion of alien life on another planet added a new dimension to Halley's visit. The effect Lowell's story had on the popular imagination was not measured, but there is no doubt it raised anxiety levels in many quarters. The universe was beginning to seem like a particularly threatening place.

The Comet Claims Another King

On the evening of May 4, 1910, London society gathered in Albert Hall to pay tribute to Commander Richard Peary, the disputed winner of the race to the North Pole. King Edward had planned to attend, but he was "indisposed."

The king's indisposition, it soon turned out, was serious. He was desperately sick. The next day, he tried to keep his official appointments, but his breathing was labored, his great body sagged. He collapsed twice and was put to bed.

• • • • • • • • • • • • • • • • •

The palace issued a typically terse medical bulletin: King Edward's condition "causes some anxiety." Nothing more had to be said, the public knew. A subdued crowd began to gather outside of Buckingham Palace.

The royal family and the highest leaders of the government also gathered, as did some of Edward's closest friends. At some point in his last hours, Queen Alexandra sent a brougham to fetch Mrs. Alice Keppel. When she arrived, the queen herself ushered her into the room where Edward lay dying, and left her alone with him.

It was May 6th. Halley had already circled the sun and was heading back out toward the edge of the solar system. In twelve days it would make its closest approach and Earth would pass through its tail. It would not be visible in England that day until three o'clock in the morning, but many eyes in the hushed crowd waiting outside the palace looked up at the sky.

Somewhere out there in space, Halley's comet was streaking toward them. And inside the palace another English king lay dying. Would he be another victim of Halley's curse?

There were many people in the crowd outside the palace, and in the rest of the British Empire as well, who deeply feared for King Edward's life. They believed his sophistication and their immense goodwill were not enough protection against the imminent danger that threatened him. After all, those worriers argued, didn't Halley's comet always presage disaster for the high and mighty? And who was higher or mightier than Edward VII, king of Great Britain and Ireland, and emperor of India? His empire, on which the sun truly never set, covered 11½ million square miles of the globe and contained over 400 million subjects, more than one quarter of Earth's entire population.

If Halley boded ill for the high and mighty, Edward indeed was in trouble. The worriers and cometophobes were convinced that the danger was real. When rational voices tried to talk them down, the fearful recalled 1066 and poor King Harold as a case in point.

King Edward died on the night of May 6, 1910. If Halley truly had a curse, it had claimed another victim. From then until the funeral two weeks later, Edward's death shared the international headlines with the two big stories of the day. One, of course, was Halley's comet. The other was ex-President Theodore Roosevelt.

When TR left office in 1909, he finally undertook the adventure he'd long been looking forward to—an African safari. After a year of hunting, specimen collecting, and writing, TR emerged from the wilds of Africa when he reached Egypt in the early spring of 1910. He originally planned to return home at once, but a series of invitations brought about what remains the most triumphal tour any American ever made of the European continent.

One after the other, the royal rulers of Europe invited Roosevelt to visit their courts. Not to be outdone, the Republic of France also extended an invitation. Roosevelt accepted as many invitations as he could possibly squeeze in and sailed for Rome, where his tour began. In every country he visited, the people welcomed him as a hero, and the court received him as if he, too, was a member of royalty. The European response clearly demonstrated that Theodore Roosevelt was without doubt the most famous American in the world.

In Oslo, Norway, Roosevelt delivered a belated Nobel speech; he'd won the Peace Prize in 1905 for his handling of the Treaty of Portsmouth, which brought about the end of the Russo-Japanese War. He was the first American to be awarded a Nobel prize in any field.

While in Norway, TR was the houseguest of King Haakon and Queen Beatrice, King Edward's daughter. When the news of Edward's death arrived, Roosevelt offered his condolences and quickly withdrew from the palace. A few days later, President Taft named him to be the official U.S. representative at Edward's funeral, scheduled for May 20, the day after Earth's encounter with Halley's tail.

The Cannons of Bermuda

Halley's part in King Edward's death quickly became a popular topic. Those who had been warning of the comet's curse now claimed proof of their predictions. But interest in the

. . . .

link between Halley and the king's death was not limited to professional cometophobes. It was particularly of interest to people throughout the British Empire. An episode that took place in Bermuda the night King Edward died was more representative than not.

Shortly after midnight, the cannons at Fort Hamilton, Bermuda, began to fire a 101 gun salute in honor of the new king. The salute took three hours to complete, so eerie explosions were still booming out over the calm waters of the bay when Halley flashed into sight at 2 A.M. It had a decidedly red tinge in its tail.

The final salute was fired at exactly 3:52 A.M., and as its lonely echo drifted out to sea, the end of Halley's tail suddenly flared up and the head glowed like a ball of red fire. This phenomenon lasted for five minutes, long enough to be seen by most Bermudans. The *New York Times* reported that "Negroes at work on the docks were overcome with terror. They fell on their knees and began to pray." Some feared it was a sign of the end of the world. Others claimed it was about King Edward's death, and was a certain sign that war or some other great calamity would happen in King George's reign. The *Times* continued: "They were speechless with fear and worked themselves up in their paroxysms of religious zeal to a perfect frenzy. It was not until the comet faded from view . . . that they could be induced to go on again with their work."

The fear shown by the Bermudans was but a faint foreshadowing of things to come.

Chapter 11

A Devil's Brew
of Cyanogen

A devil's brew of cyanogen, comet tail, a dollop
of anxiety, and a dash of reassurance were served up to the
public piping hot, in ever increasing amounts. As we've seen,
even the *New York Times* gave the comet story virtually unlimited space. Other newspapers, not so restrained, and more
eager to excite than to inform, were more lavish in their coverage. All over the world, in every language imaginable, headlines screamed at the public.

The comet-encounter stories fell into two main categories.
One was the dispute among scientists over the probable effects: would nothing at all happen, or would all life on Earth
perish? The other was the world's response: how were people
preparing for the big event?

A third aspect of the story of Earth's coming encounter with
Halley's tail involved the telling of the story itself. Perhaps
this element was seen while it was happening, perhaps not,
but with the hindsight of seventy-five years, it is one of the
most interesting elements in the whole episode. Because the
scientific community itself was split over the question of how
much danger the comet's tail actually posed, the press could
not ignore the dispute, but the way it was handled managed
to create as much doubt as reassurance.

Another destruction fantasy. The agent of doom is a comet-caused tidal wave. Drawing by Arnuld Moreaux, from The Sketch, *April 1910. (From The Library of Congress.)*

Since the earliest announcements in February, scientists on both sides of the question were quoted daily. The majority of scientists went on record in support of the nothing-will-happen theory. But the door to doom and disaster always managed to stay open, even if just a crack. In a long editorial with the oddly disturbing title, "Could the Earth Collide With a Comet," the *Scientific American* tried to offer reassurance: "It may be safely held that on May 18th next, none of us will be aware of the fact that we are literally breathing the tail of Halley's Comet." Of course, the essential element in this sentence is not the opinion that the comet's tail would be harmless, but the *fact* that Earth's inhabitants would indeed be breathing it.

As if to undermine its own attempt at reassurance, the editorial next offers a lengthy paraphrase of *The End of The World*, a recent work by Camille Flammarion, a noted French astronomer and popularizer of science: "He has shown us terrified humanity gasping for breath in its death struggle with carbon monoxide, killed off with merciful swiftness by cyanogen, and dancing joyously to an anaesthetic death, produced by the conversion of the atmosphere into nitrous oxide or dentist's 'laughing gas.' " It is true that Monsieur Flammarion went well beyond the call of scientific duty in his description of the encounter's effects, offering a variety of deaths to the public. The editorial seems to have assumed that Flammarion's ideas were so ridiculous that just to describe them was to ridicule them. Perhaps for many of the journal's readers, this was correct. But Flammarion, after all, was a scientist. The net result of the editorial probably raised doubts in some readers' minds, however sophisticated they were.

Some spokesmen closed the door to doom with finality. In Cambridge, England, Professor R. S. Ball considered the difference in mass between Earth and Halley's comet and snorted, "A rhinoceros in full charge does not fear collision with a cobweb!" Percival Lowell of Harvard turned his attention away from his Martian canals long enough to quip, "The whole tail could be packed in a suitcase." But there was Professor Henri Deslandres, director of the Astronomical Obser-

. . . .

vatory in Meudon, France, who opened the door again. What he told the world sounded like honest good sense. Science knows so little about such encounters, concluded Deslandres, "that we can neither confirm nor deny the possible dangers represented by the poisonous gases in Halley's tail."

While *Scientific American* and other apparently responsible publications addressed their words to the enlightened of the world, the popular and sensationalistic press spoke to the unenlightened, the very people who were most vulnerable to irrational fears. In the words of various newspapers, the unenlightened mass was composed of "foreigners," "immigrants," "negroes," "peasants," "savages," "heathens," "the soft-headed," "the superstitious," and "the ignorant."

The newspapers of William Randolph Hearst were particularly vivid in their coverage of the encounter. Ten days before Earth was due to cross the tail, the Sunday supplement of Hearst's many newspapers carried a double article, "Our Visitor the Comet." One part carried the headline, "Why Comets Have Caused Death, Insanity, Murder, Riots, War and Pestilence and How Superstitious Fear of Them Lingers Among Men."

The headline of the other half of this double bill told the story: "Why Halley's Comet Cannot Strike the Earth, Although *Some* Comet Will Certainly Hit Us *Some* Time—and What Will Happen Then." The author's answer? There will be no escape.

A few days later, Hearst's *Boston American* produced a mixed message that was a gem of its type: "The Earth will pass through the tail, which is composed of *masses of deadly cyanogen gas*, without any danger . . . . The brilliancy of *this ball of cyanide gas*, is the hope of astronomers and the nemesis of the superstitious." [Italics added.] After absorbing an endless stream of messages like these, what could a person think? What could a person do?

Newspaper copy was not the only source of the Halley drumbeat. The quickening rhythm came from everywhere as the everyday activities of people became entwined with the comet.

. . . .

Comet Collectibles

Entrepreneurs everywhere prepared for the comet by bringing new goods to market to meet the sudden demand. "Comet pills" and other nostrums of unknown content were widely sold. The comet pills were guaranteed to give protection against every sort of comet injury or danger. They sold well, especially in the West Indies and throughout the American south.

Telescopes and binoculars were in short supply, shops couldn't keep up with the demand. At first there was a run on the more expensive German, English, and French models, but soon there weren't even enough of the cardboard tube telescopes that sold for a dollar. Comet postcards appeared in English and many European languages. Some were amusing, others were beautiful examples of the color lithography of the day. One, for example, shows a young boy astride Halley's comet flying over the city of Pittsburgh. Another, in German, offers insurance against the end of the world.

Many of the postcards were in the German language, and Germans in general seemed to take a particularly keen interest in Halley. Perhaps their interest was heightened by the fact that a German, Wolf, had recaptured Halley in 1909, or that another German, Palitzch, had recaptured Halley on its first predicted return in 1758. German entrepreneurs produced as many kinds of comet novelties as anyone, including hats, mother-of-pearl hat pins shaped like comets, walking sticks, umbrellas, various textile goods, and several temporary brands of beer and schnapps.

Comet Cartoons

Cartoonists found Halley an unavoidable source of topical inspiration and every conceivable subject matter was tied to it: baseball, railroad fares, family life, politics, husbands who kept late hours, Theodore Roosevelt, and so on. Frequently these comet cartoons were front page items. One of them, about Roosevelt's trip to Germany, showed Kaiser Wilhelm saying, "Teddy, you're a star." to which TR answered, "Bill, you're a comet."

A widely syndicated cartoon was called "What Shall We Do With Teddy?" It reflected the nation's curiosity about Roosevelt's next move. In six separate panels, TR is shown as the Statue of Liberty; the mayor of New York; head keeper of the zoo, with the lions fleeing in fear; a baseball umpire; a U.S. senator; and as the referee for the Johnson-Jeffries fight, for which role he'd actually been considered. A seventh panel, labeled "Caretaker of the Universe," was as large as the other six combined. TR, dressed in safari gear, stands on planet Earth. On his shoulder is his famous big stick. Out in space is Halley's comet, which has just made an abrupt turn-around to avoid hitting Earth. "Pardon me," the comet says to Roosevelt, "I didn't know you were there."

Here's to Halley's

There were as many kinds of responses to the comet as there were kinds of people. On the more frivolous side, some took it as an occasion to have another drink. "Comet cocktails" were the rage of the day. Of course, the ways they were con-cocted also varied greatly. At the Plaza in Manhattan, the ingredients were kept secret, but those who drank it asked for more. They said it began with a fiery taste on the tongue, continued with a soothing sensation on the palate, and ended with a warm feeling of floating around.

A Manhattan bar featured a comet cocktail with no secrets. It was a "seething concoction made with cracked ice, a snifter of French vermouth, and a jigger of applejack." The inventor guaranteed that "six of these are enough to make a blind man see the comet."

Good-bye, Cruel Comet

While the frivolous and the profit-minded incorporated Hal-ley into their business-as-usual approach to life, the rest of the world awaited Halley with dread.

Drawing by Koek-Koek from Pearson's Magazine: U.S.A. *"If a large comet approached the Earth." A lurid example of the comet-destruction fantasies which proliferated in 1910. (From The Library of Congress.)*

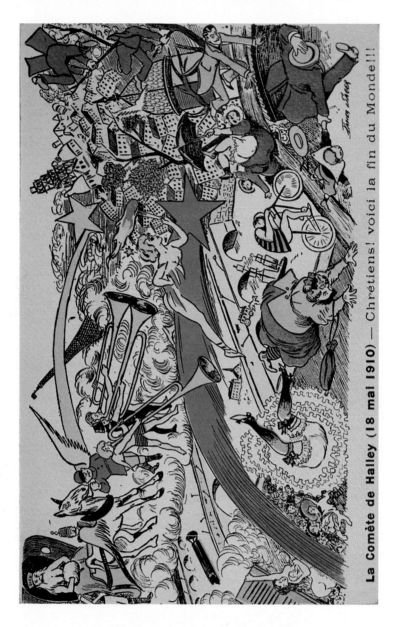

La Comète de Halley (18 mai 1910) — Chrétiens! voici la fin du Monde!!!

Postcard: France. "Halley's Comet, May 18, 1910. Christians! It's the end of the world!!!" God keeps His hands warm in a muff, angels sound the trumpets of Doom, and a curvaceous comet-nymph rides Halley as it spreads destruction. (From the M. Zwerdling–D. Miranda Postcard Collection.)

Postcard: England. The comet portrayed as a fierce and fiery bird of prey, clutching victims in its talons while others flee in panic. (From the M. Zwerdling–D. Miranda Collection.)

Illustration from The Chicago Ledger: *U.S.A. A newspaper science fiction serial starring the comet. (From The Library of Congress.)*

Cartoons by T. T. Heine from Simplicissimus: Germany. Left Panel: "The comet (looking like a dog) sees his old friends, the political functionary and the cleric, unchanged since 1835." Right panel: Professor Haekel (a well-known Darwinian biologist) sends a monkey into space so that the human race can be regenerated after the comet kills all life on Earth. (From The Library of Congress.)

Der Polizeipräsident von Jagow erläßt in seinem bekannten Plakatstil eine strenge Mahnung: „Die Milchstraße dient dem Verkehr! Kometen sind gewarnt!"

Cartoon by Olaf Gulbransson from Simplicis-
simus: Germany. The Chief of the Berlin Police
Department orders the comet off the Milky
Way (called "Milk Street" in German). (From
The Library of Congress.)

Postcard: U.S.A. The comet over Gary, Indiana. (From the M. Zwerdling–D. Miranda Postcard Collection.)

PÉRÉGRINATIONS D'UNE COMÈTE.

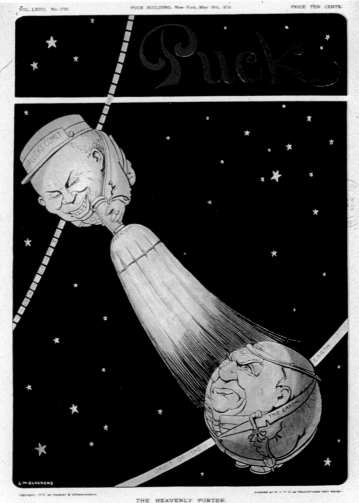

VOL. LXVII. No. 1733.　　　PUCK BUILDING. New York, May 18th, 1910.　　　PRICE TEN CENTS.

THE HEAVENLY PORTER.

Above: *Cartoon by Louis Glackens from* Puck: *England. The "Porter's brush" is reminiscent of the Chinese image of comets as "broom stars." (From The Library of Congress.)*

Opposite: *Drawing by J. Grandville from* Un Autre Monde *(Another World): France. "The wanderings of a comet." An elegant French comet fantasy, high in style and low in terror. (From The Library of Congress.)*

HARPER'S
WEEKLY

EDITED BY GEORGE HARVEY

SEE
HALLEY'S
COMET

Sidney H. Riesenberg 1910

May 21 1910   HARPER & BROTHERS, N. Y.   Price 10 Cents

Drawing by S. H. Riesenberg from Harper's Weekly: U.S.A. A news-
boy sees the comet through a "pay" telescope, one of many that
were set up on street corners all over the world.   (From The Library
of Congress.)

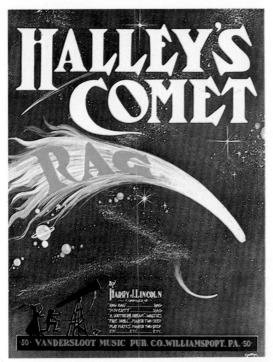

Sheet-Music Cover: U.S.A. "Halley's Comet Rag." (From The Library of Congress.)

Postcard: France. "Celestial Publicity." A gentle poke at the numerous advertisers who exploited the comet to sell their products. (From The Library of Congress.)

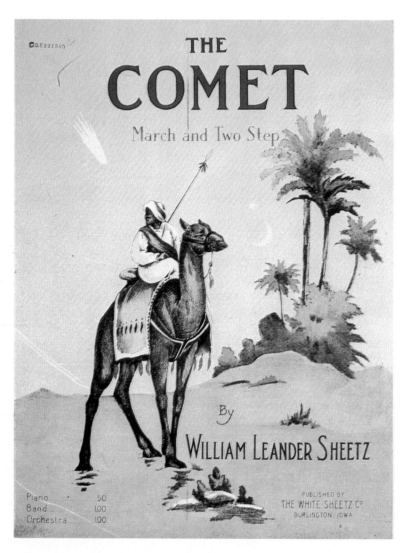

Sheet-Music Cover: U.S.A. "The Comet March and Two Step." (From The Library of Congress.)

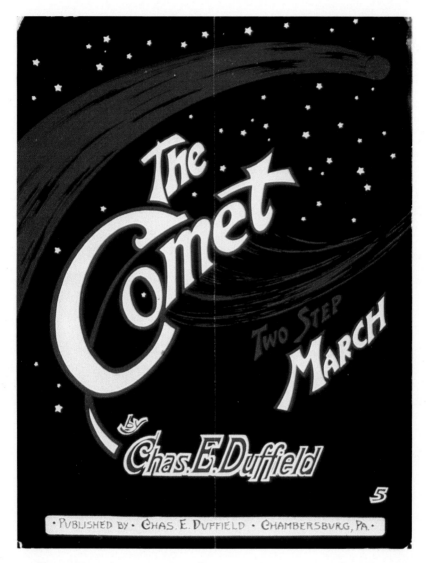

Sheet-Music Cover: U.S.A. "The Comet Two Step March." (From The Library of Congress.)

Sheet-Music Cover: U.S.A. "The Comet March & Two Step." (From The Library of Congress.)

Program for Buffalo Bill's Farewell Tour: U.S.A. This version of the
Wild West Show included the troupe of Major G. W. Lillie, known to
all as Pawnee Bill. (From the Harvard Theatre Collection.)

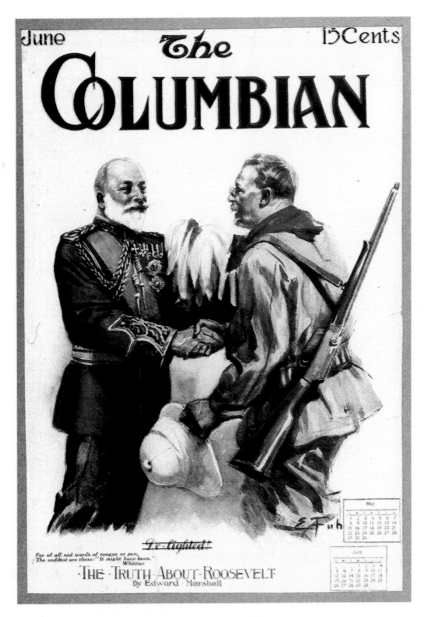

Drawing from the The Columbian: U.S.A. King Edward VII greets ex-President Theodore Roosevelt, an encounter that never occurred. The King died before Roosevelt arrived, but the magazine had gone to press. "Delighted" was crossed out and two lines by Whittier were added: "For of all sad words of tongue or pen, The Saddest are these, 'It might have been.'" (From the Theodore Roosevelt Collection, Harvard College Library.)

It is not known where or exactly when the suicides began. The *London Times* believed they started along the Mediterranean coast, and cited several examples from Spain and Italy. Southeast Europe was another locale which witnessed a wave of suicides, but this extreme response was not limited geographically. Reports drifted in from around the world.

Miss Blanche Covington of Chicago believed the comet would kill everyone in her city, herself included. To avoid the celestial suffering, she locked herself in her room and turned on the gas. A neighbor called the police and Covington was rescued. Across town, ninety-year-old Mrs. Sophie House shared the same fears, and chose the same solution, but she was successful. She killed herself by inhaling illuminating gas.

In California, another fearful person chose suicide, but he did it with a good deed in mind. Paul Hammerton, described as a "brooding, melancholy type," was a prospector in the San Bernardino Mountains. He was convinced the comet's would end all life, but he thought the disaster was avoidable. He believed he could save the world if he could think of the right solution. He brooded on it, but said nothing. Then the answer came to him.

Early one morning, a few days before the encounter was due, Hammerton set out from town and went up into the hills. He knew what he had to do. When he reached the site of his mining claim, he gathered some timbers and erected a crude, life-size crucifix. Then he took his hammer, a few long spikes, and methodically nailed himself to the cross.

First he nailed his left foot, then his right. He pounded the spikes through his flesh and into the wood. Then he reached over and hammered a spike through his left hand. Now it was done. He could go no further. He dropped his hammer and hung from the cross.

It is not certain how long Hammerton endured his crucifixion. He was still conscious when some fellow prospectors discovered him. He was in intense agony, but he pleaded with his rescuers to let him stay where he was. It was for their good, he told them, what he was doing would save the world

. . . .

from Halley's comet. But his friends ignored his pleas, un-nailed him, and carried him down to San Bernardino. He was treated and eventually declared insane.

Some of Hammerton's fellow cometaphobes remained convinced that his crucifixion, incomplete as it was, had served its purpose. After all, they pointed out, the world was saved, wasn't it? Cynics simply shrugged and said he was trying to carry self-sacrifice too far.

Comet Crazy

Poor Mr. Hammerton was not the only person driven crazy by fear of Halley. Cases of insanity attributed to the comet were reported from many countries. In accordance with the general English view of southern Europeans, the *London Times* declared that southern Europe was having more comet insanity than anywhere else. But there were just as many cases on the other side of the Atlantic. In Orrville, Ohio, for example, Fred Bowers became "insane from brooding over Halley's comet."

For reasons that are not altogether clear, the people of Chicago were particularly upset. A woman from the North Side was riding a streetcar downtown when suddenly she rose to her feet, shouting and gesticulating wildly. "Glory, glory, glory!" she cried. "Get down on your knees, you sinners, and pray the Lord to forgive you, for this is the end of the world." She tried to make everyone on the car get down on the floor and pray, but a policeman boarded the car and took her away. In another part of Chicago, Sam Popowski loudly complained that the tail of the comet kept hitting him and was beating him to shreds. A city physician pronounced him insane.

In New York City a well-dressed woman was observed standing at rigid attention on the steps of City Hall. After a while, a policeman asked her if anything was wrong. "I'm waiting to see the Judge Mayor," she replied in a normal voice. "I must tell him many terrible things about Halley's comet." The policeman took her to the Elizabeth Street station, from which they sent her to Bellevue Hospital for observation.

Preparing for Halley

As the night of the encounter drew closer, millions were frightened into prayer. Churches were never empty. In some parts of the world, a lightning bolt, a thunder peal, a photographer's flash lamp, even a loud backfire was enough to drop whole crowds to their knees praying fiercely.

For millions of others with a more practical turn of mind, fear led to thoughts of preservation. In urban areas, people tended to believe they could survive the tail if they could make their homes airtight. Consequently, windows, chimneys, and later on doors, were sealed up all over the world.

In the rural areas, folks tended to dig in. Most used root cellars, and in the American midwest and southwest, cyclone cellars were the shelters of choice. A few had to build new. In Council Bluffs, Iowa, farmer John Marlow dug a cave for his family. Power tools were not yet available, he had to use a hand shovel. He must have worked hard and long. He even built an air-tight door. By the time he was done, his comet shelter was big enough to hold him, his wife, his children, two horses, two cows, a cat, a dog, and a number of chickens, and he herded them all in. Somehow, the Marlows survived.

A young office boy on Wall Street boasted he had the best solution. He had already collected a few long hollow reeds. When the tail entered the atmosphere, he was going to duck underwater at the edge of the East River and breathe through a reed. He wasn't sure why his boss laughed when he told him his solution.

A resident of the Waldorf-Astoria followed the comet stories closely. A few days before the encounter, she hurriedly left New York and headed for her lodge in the Adirondack Mountains. If she was going to die, she announced, she preferred to do so in the mountains rather than in New York City.

Very elaborate attempts to escape the comet's tail were carried out. The sale of bottled oxygen suddenly got very brisk. The people who bought them worked on the theory that the tail would pass through the atmosphere rather quickly, and their store of oxygen would hold them over until it passed. In

several European cities, newspaper ads appeared announcing
the availability of underground brick chambers, equipped
with bottled oxygen. It was reported that such chambers had
been built at the bottom of some South African mines for the
use of management.

Eyes to the Sky

As the world's response to the coming encounter with the tail
intensified, the newspapers continued their barrage. Professor
A said "No danger at all," immediately followed by Professor
B, who warned "Anything can happen."

Fearful or not, everyone wanted to see the comet. The hours
of its visibility were announced in every news edition. For a
good part of its visit, Halley was visible as a morning star, and
a very early morning star at that. So all over the world people
got up early, or stayed up all night, just to get a glimpse of the
famous visitor. Crowds gathered on Hampstead Heath near
London, along the Charles River in Boston, on Capitoline Hill
in Rome, on rooftops, in open fields, at the seashore, and on
any other spot that afforded a good view. Many of the watchers formed "Comet Clubs," which flourished for a few weeks
and then disappeared.

Sightseers and astronomers ascended into the skies in hot
air balloons to get a better view. One astronomer and his wife,
on a balloon flight in New England, were fired upon by a
rifleman. The incident was attributed to fear. The man with
the rifle was believed to be a farmer who looked up at the sky,
saw a strange round shape with a little fire beneath it, and
assumed he was being attacked by the comet. Luckily for the
astronomer and his wife, the unknown farmer's aim was poor.

In the suburbs of Boston, a young couple had a comet en-
gagement party, held outdoors in the early hours of the morning. Telescopes were provided for the guests, and a Harvard
astronomer gave a little talk.

In a letter to the editor of the *New York Times*, one wag
wondered if our contact with the tail would render us all
"comet-ose."

The Comet Is a Woman

At an elegant after-theater comet supper, a Chicago matron announced she had no fear at all of the comet. She believed it was one of her own. "The comet is a woman," said Mrs. D. F. Burke. "Men sit up all night to look at her; she is brightly attractive and possessed of charms for all observers; she is brilliant, dangerous; she moves the heavens its wonders to perform; she is not answerable to petty laws; and she is absolutely necessary in the scheme of things."

The responses to Halley's arrival were varied, but the intense anticipation of the coming event was universal. Whatever comet antics people performed, they were part of a great whole. Schoolchildren wrote stories and drew pictures about it. Housewives spoke of it over tea or on their shopping tours. Businessmen brought it up in every conversation. Police were harried by comet pranks. Comet dinner parties and comet dances were the fad of the day. The single cause of all the action was Halley's comet: beyond any doubt, it had captured the imagination of the entire human race.

A comic version of genuine fears. (From the Zwerdling–Miranda Collection.)

Chapter 12

"Six Hours in Comet's Tail Tonight!"*

The night of May 18–19, 1910, witnessed what is arguably the largest simultaneously shared event in history. Without doubt, it was certainly the largest such event before the age of instant, universal communication. For one thing, word of the comet's arrival did not have to be broadcast, all that people had to do was look up in the sky. As for the controversy about Earth's passage through the cyanogen tail, this message managed to reach everywhere because of the prolonged and incessant newspaper coverage. By the time the news collection was over, reports had come in from virtually every country on Earth.

The Disappearing Dogs of Constantinople

By the standards of Western enlightenment, the city of Constantinople did not belong to another age, it belonged to another world. By long standing custom, the small boys of Constantinople were dressed like girls to disguise them from the devils. The assumption was that girls were so worthless

*Headline, the New York Times, May 18, 1910.

• • • • • • • • • • • • • •

the devils wouldn't even bother with them. All beasts of burden and many people wore charms and amulets to protect them from the evil eye. An event as predictable as an eclipse was enough to throw most of the population into a state of abject terror. It was no wonder that the night of the comet's tail was a night of fearful expectation in the city on the Golden Horn. Reported the London Times, "The predictions of impending disaster . . . produced a highly nervous state among the superstitious and ignorant population in Constantinople."

Many families felt the need to be together if the end came, so parents kept their children home from school. Children who did not go to school were gathered up from the streets and brought home by anxious parents. The streets of the ancient city were quieter than usual. Even the city's well-known street dogs seemed to sense something in the air, and moved deeper into the shadows than usual.

Few people went to bed that night. Instead, by perhaps the hundreds of thousands, they flocked to the flat rooftops of the city and watched. As the night wore on, children began to fall asleep. To stay awake and bolster their spirits, the grown-ups now and then clapped their hands. The sound of clapping floated from rooftop to rooftop all over the waiting city.

The comet's tail was invisible close up and it had no taste or odor. The only way the fearful would know that the tail had arrived was if their fears came true and they died. So in Constantinople, just as everywhere else, there was nothing about Earth's encounter with the tail to actually see. When dawn arose over the minarets of the city, nothing seemed to have happened. The sporadic clapping gave way to a sudden wave of loud cheers. The few who had gone to sleep woke up complaining, but the happy throngs up on the rooftops paid no heed.

At first everything seemed the same, but when the crowds came down from the roofs and poured into the streets, they saw that something had changed. The dogs were different, there were fewer of them. In some streets, they had completely disappeared.

. . . .

The street dogs of Constantinople were notorious. There were thousands of them. Generally, they were friendly and well fed. The citizens were quite fond of them, but almost none would take them into their homes. The reason was religious law; the Koran declares that dogs are unclean animals. The problem was, how to control the dogs?

It was an old and complex problem. On the one hand, the citizens were genuinely fond of the animals. They fed them, and some people even bequeathed money in their wills for the care of street dogs. But the citizens maintained the holy law and would not take the dogs home. In the nineteenth century, the Sultan Abdul Medjid once exiled the dogs to the Island of Marmora. But it soon became apparent that the dogs were needed for their services as scavengers; they helped to keep the streets clean. And more importantly, the public was upset. The citizens raised such a clamor on behalf of the dogs that the sultan relented and brought them back to the city. The people gave them a great feast of welcome.

In 1910, Turkey had a new regime. The municipal authorities of Constantinople, more forward thinking and modern than their predecessors, viewed the dogs as a public health menace and a general nuisance. They wanted to remove them, but they recalled the experience of Sultan Abdul Medjid and feared a protest from the indulgent public. What to do?

Then a particularly shrewd official had an idea. Why not wait for the night of the eighteenth to round them up? He reasoned that the work could be done without interference that night because the population would have its mind on something else. The idea was accepted. While a good portion of the city's population was up on the rooftops looking at the sky above, the great dog roundup took place on the streets below.

Squads of policemen and streetsweepers did the work. Armed with lassos or long-handled wooden tongs, they quietly worked their way through the city, capturing every dog in sight. The dogs, trusting creatures that they were, offered little resistance. They were unceremoniously dumped into dust wagons and wheeled off to holding pounds.

• • • • • • • • • • • • • • • • •

When the citizens discovered the dogs were missing, they raised a great clamor and rushed off to the holding pounds. The victorious authorities refused to release the dogs en masse, but they did allow a few to go free if a citizen paid a bail fee and promised to buy a dog license as soon as they became available. As for the unbailed dogs, they simply disappeared.

The fate of the missing dogs of Constantinople was left to speculation. One source said they'd been marooned on an island. Another said they were being held in canine "concentration camps and fed on municipal rations." But the strongest rumor, the one that got most repeated, was also the most practical: it simply said the dogs had been made into gloves.

Paris, France

All of Paris was prepared for comet night. Cafés stocked up on supplies, churches remained open, peddlers of comet novelties manned their stalls. The evening began with an extraordinary thunder shower, which added another ominous note to the city's mood. When the rain stopped, the sky remained overcast and a sense of disappointment prevailed.

As Parisians are known to do, they gathered in the cafés and lightened their disappointment with wine and other restoratives. Montmartre restaurants and cafés were as crowded as on the annual celebration of Reveillon. Everyone drank, and everyone had something to say about the evening's events. Balloonists cursed the weather; it had prevented their ascents. Between drinks, the fearful goaded each other with dire predictions that something terrible was really going to happen. Those who merely wished to see the comet on this special night wondered if the clouds would lift in time.

French astronomers had predicted that Earth would enter the comet's tail sometime after 3 A.M., though there was some dispute about the time. A few scientists were even claiming that there would be no contact with the tail at all. The uncertainty of the scientific community only added to the anxiety of the population in general. By late afternoon, people began

. . . .

to fill the churches. Throughout the city, others tried to make their apartments air-tight. And like their colleagues around the world, the super-cautious had obtained bottled oxygen and were passing the fearful night underground. A lucky few took cover in well-stocked wine cellars.

Sometime in the evening, crowds began to gather in the open places of Paris, particularly along the banks of the Seine. Maybe the cloud cover would lift and they would be able to see something. As the evening wore on, some grew more anxious, some knelt in prayer, others carried on as if they were at a gigantic nighttime picnic.

"Comet suppers" were featured at the elegant restaurants of the city, and "comet balls" were attended by the smart set. Gentlemen were requested to wear pale blue evening dress to match the blue hues of the women's gowns. The blue clothing was supposed to represent the sky.

The mixture of gaiety and fear was evenly divided in Paris, but in the rest of France, fear prevailed, particularly in the small villages and in the countryside. In the south, several more suicides were added to the list of Halley's victims. In Lille, strange pink and violet lights were seen in the sky, adding further to the widespread terror.

Atop the Eiffel Tower, the highest structure in the world, a group of astronomers stood their damp vigil, hoping against hope that the sky would clear in time to give them a brief view of Halley as it passed across the sun. But conditions did not improve, and they waited in vain.

By morning, Paris and all of France returned to normal.

New York, New York

Halley's comet began its New York run on May 10. It was visible between three and four in the morning, but the wait had been worthwhile. Its slow passage across the sky, moving at about the same rate as a planet, was awesome. Its glowing tail stretched for 25 million miles. Everywhere in the city people got up early, or stayed up late, to see it.

The sale of comet novelties peaked, and telescopes or bin-

. . . .

oculars were almost impossible to find. Comet events—parties, suppers, dances, dress balls, breakfasts—proliferated. Some of the most amusing were given by members of the smart set, who favored the rooftops of the city's hotels. For a week or more, a comet party of some kind was held on almost every Manhattan hotel roof. The demand for roof space was so great that many hotels insisted on reservations for visits to their rooftops, particularly for May 18, the night of the encounter.

On the eighteenth, all but two of the major league teams played games. From the point of view of Halley's comet, the most interesting game of the day was played at Hilltop Park (also known as American League Park) in the vicinity of 168th Street in New York. Hilltop Park, so shabby that it had to be abandoned a couple of years later, was the home of the New York Yankees. That day they played the lowly St. Louis Browns. The game itself is less interesting than the account of it that appeared next day in the *New York Times*. The article is unsigned, few sportswriters were given bylines, but it was probably the work of Dan Daniels. It is a little tour de force:

> "Comet day was observed at American League Park yesterday with appropriate astronomical fireworks, the Yankees gushing forth a shower of meteorites in the seventh and eighth innings which put the St. Louis Browns groggy under the strong influence of cyanogen gas. As the Hilltop is high, a lot of the fans thought it likely that this streamer of fire that young Halley is driving was likely to strike there, and stayed away. Those who did observe the display liked the pyrotechnics. The Yankees won, 6 to 3."

The long newspaper barrage had prepared New Yorkers, and the appearance of the comet in all its glory overwhelmed them. By the millions, New Yorkers came out to observe the night Earth encountered Halley's tail. The wide range of their activities reflected their great diversity.

Crowds began to gather early in the evening. Favorite spots were Central Park, Riverside Drive, Grand Street on the Lower

.   .   .   .

East Side, various intersections in Greenwich Village, bridges, and most of the city's rooftops.

Then as now, New York was a polyglot city. But the foreign-speaking citizens were not deprived of the Halley news, the numerous foreign language newspapers saw to that. The Yiddish, Russian, and Polish dailies kept the East Side informed, the Italian dailies informed Greenwich Village and other Italian speaking enclaves.

At eight o'clock in the evening an eerie procession filed by the old St. Patrick's Cathedral on Mott Street on the Lower East Side. Hundreds of Italian schoolchildren, dressed all in white and led by a tall man dressed in his Sunday best, marched down the middle of the street chanting the litany of the Blessed Virgin. A large crowd watched from the sidewalks. The policeman on duty in the neighborhood asked the tall man what it was all about, but the man silently brushed him aside and gestured up at the sky. A short time later other policemen arrived on the scene and ordered the people off the streets.

A large group of Italian-speaking citizens gathered at the intersection of Broome and Mulberry Streets to watch for the comet. The dominant mood was apprehension and the crowd was hushed. Suddenly a ball of fire appeared in the sky. Many in the crowd fell to their knees and began to scream in Italian, "La fina del mondo!" The noise brought more people rushing from their houses. With the crowd now swelled to several hundred people, the fireball burst into a shower of fire droplets. Panic swept the street. Terrified voices prayed for deliverance, and frantic pleas were made to special patron saints.

A couple of skeptics in the crowd examined the fire droplets as they fell to the sidewalk and saw that they were only burning pieces of rubber balloon. As they loudly teased their frightened neighbors, the police arrived and dispersed the crowd. The pranksters who sent the balloon aloft were not discovered.

The *New York Times* reported that on the eighteenth, "many foreigners living at Constable Hook, Bergen Hook and Bergen Point . . . refused to work, and many spent the day

. . . .

offering prayers in the churches for deliverance." Many large manufacturing and construction firms were so understaffed they had to close down for the day.

The Lower East Side was one of the liveliest areas of the city that night. The rooftops were full and the streets were packed, particularly Grand Street, a main thoroughfare, where the crowds were so dense it was difficult to pass. The air was full of many languages—Yiddish, Polish, Russian, Ukranian—as people spoke excitedly about the comet.

Late in the evening, some boys on a Grand Street rooftop released a balloon with a red light attached to it. It shot up into the sky and was immediately spotted by the huge crowd. Wild confusion followed, punctuated with shrieks and prayers in many languages.

The wind carried the fiery red balloon eastward, and in a few moments it was visible from the Williamsburg Bridge, on which another 20,000 people had gathered to watch for the comet. "There it is!" someone screamed, and another panic was on. Before the balloon sailed out of sight, the prank was discovered, the crowd laughed at its own gullibility and settled back to waiting for the real thing.

Riverside Drive, with the open space of the Hudson River below, was one of the favorite spots that night. The crowds stretched out for block after block. An early evening rain shower scattered the watchers, but when it was over they returned.

Across the Hudson River, in Plainfield, New Jersey, another prankster went to work. A searchlight manufacturer mounted a battleship-size searchlight on the roof of his factory. At dusk, he sent a circle of light racing across the undersides of the clouds. His homemade comet terrified many people in the area, and thrilled others.

In Woodbury, New Jersey, Mayor Ladd offered a special service to those of his constituents who wanted to see the comet but did not want to stay awake all night waiting. The mayor arranged a comet alert. He instructed the police chief to be on the lookout for the comet and to telephone those slumbering citizens who wanted to see it. When the comet came into view early in the morning, a few dozen households were awakened by a short phone call from the local desk sergeant: "Get up and see the comet!"

At 10 P.M., Professor Harold Jacoby of the Columbia University Astronomy Department received a frantic telephone call from a man in Larchmont. "I can see it!" the agitated caller shouted. "It's on the eastern horizon and it's blinking!" Professor Jacoby calmed his caller by gently telling him he was looking at the off-shore lighthouses, which had been there for many years.

The rooftops of the city's apartment houses were crowded. Some people talked and joked, some prayed, many were silent. On such a rooftop in Brooklyn, a group of young people had gathered. Their elders were on rooftops nearby. At first the youngsters were full of excited talk, but as the night wore on, their voices fell silent. Now and again a word of reassurance would drift from one rooftop to the next. The youngsters stretched out on the roof, or leaned against chimneys and waited. Suddenly the quiet was broken by a scream and the sound of breaking glass. Sixteen-year-old Amy Hopkins, apparently moving to a better spot, had crashed through a skylight. She fell four floors. When the ambulance arrived, she was pronounced dead on the spot. Her mother, on the rooftop across the street, came down to see what the commotion was about. Her neighbors tried to hold her back, but she broke through. When she saw her daughter's body, she fell in a dead faint.

Thousands of New Yorkers thronged to Central Park. Their wait was marked by the usual pranks and rumors, but they were not allowed to wait until Earth encountered the comet's tail at around three o'clock in the morning. Starting at midnight, a wave of policemen marched across the open fields, clearing the park of people. At the 59th Street exit alone, over 5000 people, trying to leave all at once, got hopelessly snarled in a pedestrian traffic jam.

High above the turmoil of the streets, the smart set gathered on the rooftops of their favorite hotels. The hotel managers, swamped for more than a week with rooftop requests, set up

· · · · · · · · · · · · · · · · · ·

special rules. Some required reservations for nonresidents who wanted to observe comet night on their roof. Many set up special couches, or corners filled with pillows, to afford the best and most comfortable view of the sky. Several managers installed telescopes. The Plaza installed a temporary wireless telegraph station on its roof, so its guests could learn the latest comet news from around the world.

On the Waldorf-Astoria there was a brief panic. While several hundred men and women were scanning the sky, a photographer took a flash picture. The sudden and unexpected burst of light, thought by some to be the "crack of doom," set off a round of shrieks and screams. The panic quieted down when the photographer explained what had happened.

The crowd on the Waldorf was also treated to the presence of "Uncle Joe" Cannon, Speaker of the U.S. House of Representatives and one of the most powerful men in the country. He seemed delighted by the whole event, and grew more animated when several of his good friends appeared in a large contingent from the National Association of Manufacturers, which was holding a convention in the hotel.

The crowd grew boisterous at times, but one small group on the Waldorf carried on as if it was a normal night. A group of women in the glass-enclosed sun parlor played bridge until the early hours of the morning, and not one of them was seen to look up from her cards.

The Netherlands, the Majestic, the St. Regis, the Astor, the Knickerbocker, the Hotel Manhattan, the Belmont—just about every major hotel in Manhattan had its rooftop comet party. Some were specially decorated for the event. The Gotham, for some strange reason, chose to erect wigwams and tepees, though nobody could understand why.

At all the hotels, guests sipped comet cocktails, nibbled comet hors d'oeuvres, ate comet suppers. By early morning, when it was seen that the comet's tail had caused no damage, comet breakfasts were served.

.   .   .   .

Dateline: The World

- San Juan, Puerto Rico: So many workers failed to show up that plantations, tobacco factories, and pineapple loading docks closed down for the day. Churches were packed all day and night. Long processions wound through the island's cities—men, women, and children carrying candles and chanting.

- Wilkes-Barre, Pennsylvania: Mine workers in the anthracite region refused to enter the mines. They said they wanted to be above ground to meet the end if it came. Many prayed all day. Several collieries closed for the day.

- On board the liner "Germania," in mid-Atlantic: The 370 steerage passengers passed the night on deck. The comet was clearly visible. Some stared at it, brandishing crucifixes, but others turned away and would not look. Many knelt on the deck and prayed. Their great wish was to reach land before the comet did its worst.

- Mexico City, Mexico: Huge throngs gathered around crucifixes set up on hillsides. For ten days, "superstitious" Mexicans had sought to avert the world's end with incantations, music, prayer, and what were called "weird ceremonies." On the night of the encounter hardly a soul remained indoors. By dawn, when the danger seemed to have passed, the mighty crowds celebrated with feasts and dancing.

- Prague, Czechoslovakia: Franz Kafka spent part of the night on a rooftop with his friends. Later he wrote in his diary, concerning the fact that he had not been writing much lately, "But everyday at least one line should be trained on me, as they now train telescopes on the comet."

- Lausanne, Switzerland: The Swiss Aero Club arranged for balloon ascents. Alpine resorts throughout the country were full of comet watchers. Midnight comet dances were held in many cities; hotels and restaurants stayed open all night.

- Pitcher, New York: The prize Dorsetshire cow of dairy-man Amos Rhodes gave birth to four well-formed calves. Two of them had distinct star-shaped markings on their foreheads. The celestial birthmarks were immediately attributed to the influence of the comet.
- St. Petersburg, Russia: Fear was widespread among the lower classes, the churches were open all night. While the fearful prayed, epicureans partook of elegant comet suppers in the finest restaurants, followed by rooftop comet dances.
- Talladega, Alabama: Miss Ruth Jordan, a white woman, was called to the door to see the comet. She looked up and dropped dead. Not far away, at the railroad depot, an unidentified black man was shown the comet. He dropped dead too.
- Washington, D.C., the White House: President and Mrs. Taft, curious as the rest of the world, spent several hours peering out of the White House windows. The sky was overcast, and they had no luck.
- A mountain village, Hungary: The entire population of a small village was convinced "the end" was at hand. After praying in their church all day, the people rushed back to their houses to prepare for the dreadful night. Doors, windows, and chimneys were stuffed with rags to keep the cyanogen out. All the villagers huddled indoors except for one, the stalwart night watchman. He felt duty bound to stand his watch, even if it was the last night of life on Earth. He, too, had no doubt about the night's outcome, he just felt obliged to be the first to spot it.

At three in the morning, as Halley slowly floated overhead, the watchman glanced across the valley and saw a small fire in the next village. It was all the proof he needed that the end had begun. He raced through the village, alternately blowing on his loud horn and shouting, "It's the end of the world!"

Everyone poured out into the narrow streets. The women and children came first, shrieking and bawling in terror. The men followed. They, too, were filled with fear,

. . . .

but they managed to remain practical. After a moment of quick discussion, the men suddenly raced back into their houses and returned with all the food and beverage they could carry. The whole town rushed to the small square in front of the church and quickly lit a great bonfire. In a few minutes the night was filled with the aroma of a huge barbecue. Wine bottles were freely passed around. Everyone joined in the grandest feast the village had ever seen. In between mouthfuls, the people glanced up at the comet and offered brief prayers for survival.

- Ohio, U.S.A.: Humorist James Thurber recalled that night in an autobiographical reflection: "I remember when the scientists were not so cheerful; I remember when Man and his flimsy globe were doomed. I had become, before the good news arrived, a timid eschatologist, quietly waiting for the finish of this fragile Luna moth we call mortality. In 1910, when I was a stripling of 16, the bearded watchers of the sky (at least those in the Middle West) fondly predicted that Halley's comet was going to strike the planet. Nothing happened, except that I was left with a curious twitching of my left ear after sundown and a tendency to break into a dog-trot at the striking of a match or the flashing of a lantern."

- Berlin, Germany: Thousands of citizens attended all-night picnics on Kreuzberg Hill or high in the forest of Grunewald. Others took all-night steamship excursions the better to see Halley. Open-air restaurants and beer gardens all over the city stayed open all night. Some people remained in the safety of their sealed-up houses. Others went to church and prayed all night.

- In Great Britain, the population faced the night of the comet with mixed emotions. Due to the king's recent death, the response to Halley was more or less subdued. Many Londoners stayed at home, trying to seal up all the cracks to keep out the poisonous gas. Others retired to cellars with containers of bottled oxygen. But by nightfall, the streets of London and other English cities were filled with people. Some huddled in quiet little groups, other carried on boisterously as if they were enjoying a revel. The night was hazy, but Halley was not due to appear until 3 A.M. Perhaps the haze would lift.

At the hour of encounter, the sky remained hazy. The revelers paused. In the silence, prayers could be heard. But by 4 A.M., when nothing seemed to have changed, the mood of the crowds grew festive. The air was filled with cheers, with sighs, and people began to drift off toward their homes. The night of the comet's tail had come and gone: now England could prepare for King Edward's funeral.

The Great Funeral took place on Friday, May twentieth. Even the most fearful of the cometophobes could not have imagined that, with the comet overhead, they were about to witness not merely a funeral procession, but the end of an era. Edward's funeral was to be the last great gathering of the crowned heads of Europe and the first historic European moment which featured an American, Theodore Roosevelt. It was a symbolic farewell to the old order of the nineteenth century and an early glimpse of the new. Afterward, there would be a nameless four-year period, a kind of treading water as the world drifted into World War I and the start of the modern age.

Chapter 13

Was Halley Offered a Virgin?

Earth had passed through the comet's tail and nothing had changed. We were not dead, as so many had feared. A great weight had been lifted from the public's imagination. Now the world was free to enjoy the spectacle of Halley, and indeed it was more visible and gaudy on its outward passage than it had been before. But comet fever was not eliminated, it was only dormant. It lurked in the corners of the human imagination, waiting.

Now came one of the most bizarre comet tales of all. In the May 20th edition of the *Oklahoma City Times*, the following article appeared:

Girl Rescued from Death at Gory Stake

ALINE, Okla., May 19.—Jane Warfield, a pretty nineteen-year-old farmer girl, living near here was rescued after a hand-to-hand conflict between members of the sheriff of Alfalfa county posse and Henry Heinman's religious fanatics Wednesday evening just as the girl was about to be offered as a blood sacrifice for the atonement of the world's sins in order that Halley's comet might not destroy the earth.

The girl, nude and partially unconscious was tied to a stake in

GIRL RESCUED
FROM DEATH
AT GORY
STAKE

ALINE, Okla., May 19.—Jane Warfield, a pretty nineteen-year-old farmer girl, living near here was rescued after a hand-to-hand conflict between members of the sheriff of Alfalfa county posse and Henry Heinman's religious fanatics Wednesday evening just as the girl was about to be offered as a blood sacrifice for the atonement of the world's sins in order that Halley's comet might not destroy the earth.

The girl, nude and partially unconscious was tied to a stake in

A news item from the Oklahoma City Times, May 20, 1910. (Photographed by Donald Gropman.)

the center of a dancing group of the crazed followers of Heinman and within a few minutes was to have been stabbed and bled to death. Heinman's chief prophet was ready to perform the deed.

It was known in the community that the much-heralded approach of Halley's comet and that threatened danger attached to its appearance had affected the fanatics and frequent meetings were being held. All their secrets are closely guarded and it was not until the girl was tied to the stake that the authorities became aware of the intended sacrifice.

Posse Starts Out

A posse was immediately formed and proceeding to the meeting ground of the fanatics the girl was rescued and given medical attention. Followers of Heinman attempted to fight the officers, but they were overcome with little difficulty. Heinman was arrested and placed in the county jail.

Heinman instigated the act by telling his companions that the comet meant the end of the world and the sacrifice was necessary for their atonement.

We don't usually associate the practice of human sacrifice with twentieth-century America. The story sounds like it came out of the Dark Ages. But there it is, in black and white, datelined Oklahoma on the night of the comet's tail. Its terseness is maddening. Who were these people? What did they do and why? How did the story end?

Perhaps we can imagine the terrified victim, half-crazed from fear. Or perhaps she was numb, shocked into dull lethargy by the unimaginable thing that was happening to her, hardly aware of the ropes that bound her to the stake.

And there is the unnamed chief prophet, knife in hand, about to pierce the flesh and start the flow of blood. The frenzied followers dance in a circle, ecstasy on their flushed faces, eager for the ritual to reach its climax.

Finally, Heinman himself, standing beside the chief prophet, swaying back and forth chanting his incantation. When the posse suddenly breaks over the ridge, he orders the chief prophet to act now, immediately, to have done with it before the posse arrives. And then the struggle, and the res-

• • • • • • • • • • • • • • • • •

cue, and the courtroom scene in which Jane accuses Heinman to his face, and he, filled with righteous wrath, thunders his own mad defense in the name of Halley's comet, which must sooner or later be appeased with virgin blood! As the camera draws back, we see the mad Heinman being led off to prison and Jane, in the arms of the sheriff who saved her, looks up into his ruggedly handsome face and says with a faint smile, "Thank God it's over at last!" and he replies, "Now it's time to see the parson."

It sounds like a movie, and it could be a terrific one at that, sort of a gothic-western in which the good deputies battle the evil cultists for the life of the young maiden. And a movie is most likely the only way we will ever hear the end of this strange story, for in all likelihood, it was a hoax.

The story of the Oklahoma comet cult has been retold in several books, at least one of them in French. For the most part, the authors of these books are sophisticated writers, but they included the cult story as if it was a true event. There is no interest here to quibble with their research. But what is of interest is the fact that these writers should have accepted the story as the literal truth. They assumed that the hold comet fever had on the human imagination could have been strong enough to make the story entirely plausible.

In the research for this book, the discovery of the *Oklahoma City Times* article was the high-water mark in the comet cult department. From there on, the tide ran out. No follow-up story was ever discovered, in this or other newspapers. Members of sheriffs' offices in three different Oklahoma counties searched their old records for any mention of the affair, but found none. Court records also revealed nothing. Several very tolerant Oklahomans named Warfield were interviewed, with no result.

Others who assisted in the effort to pin down this story include three experts on Oklahoma history and folklore, the record office of the Oklahoma Prison System, staff members of the Western History Collections at the University of Oklahoma at Norman, and most fortuitously, Eleanor Landon, of the Oklahoma Historical Society, who discovered the *Oklahoma City Times* article in the first place.

Was Halley Offered a Virgin?

In 1984, a few Oklahoma newspapers printed stories about the search for information about the cult. One was the *Oklahoma City Times* itself, in which Kent Ruth, who writes a widely read column called "Windows on the Past," did a piece entitled "Sacrificial Virgin Saved Just in Time, Story Says." No readers ever came forth with any information.

Everything seems to point toward a hoax, but there is one curious fact that causes a glimmer of doubt. The two people named in the article, Jane Warfield and Henry Heinman, both appear on the 1910 Oklahoma Census list. They were real people. If their names were used in a hoax, why was there no retraction? Why no lawsuit for defamation of character? And most baffling of all, why has nobody from the area ever heard of the story?

Unless somebody comes forth with definite information, one way or the other, we shall never know the truth.

Part Four

Our Rendezvous with Halley

Marching down Fifth Avenue for Women's Rights. (The Bettman Archive, Inc.)

Chapter 14

The More Things Change...

Our generation has a rendezvous with Halley in 1985–1986. For most of us, it will be a once-in-a-lifetime experience—how many who are here now will be alive to see Halley return in 2061 A.D.? But for Halley, Earth is a familiar waystation. It has flown by perhaps fifty or sixty times. For us, each visit marks a lifespan, but how much change actually occurs between visits?

In recent years, futurologists such as Alvin Toffler have repeatedly told us that the rate at which things change has itself been increasing at an unbelievably fast pace. In technology, for example, this seems to be an obvious truth. But in many other facets of life here on Earth, the rate of change is all but invisible. Some of the issues that raged during Halley's 1910 visit seem hardly to have changed at all.

Latin America

In the spring of 1910 the Panama Canal was still under construction. Newspapers periodically updated the progress of the monumental work. In May, over 3 million cubic yards of earth were excavated.

. . . .

Meanwhile, our other involvements in Latin America went on apace. The United Fruit Company established its own wireless system. America's political involvement in the area was sizzling. By the end of May, U.S. Navy gunboats were shelling the town of Bluefields, Nicaragua. A front page story in the *New York Times* has an ominous sense of déja-vu: "Battle Raging. Hundreds Slain. American Gunners Beat Back Attack of Nicaraguan Army and Save Estrada from Total Defeat. Warship Shells City."

Drug Abuse

Drug abuse was a major national problem. Alcohol was the most commonly abused drug, and the legendary prohibitionist, Carry Nation, was still on the scene, though her days of smashing Kansas saloons with her axe were in the past.

The prohibition of alcohol was an issue in many parts of the country, and many individual counties were already dry, though national prohibition would not begin for another nine years with the passage of the Volstead Act in 1919.

Marijuana was still legal, and its most common use was in patent and prescription medicines. In fact, it was one of the most widely used drugs in the American pharmacopoeia.

Cocaine, on the federal level, was also legal, though here, too, some states had taken action. In the words of the *World Almanac and Encyclopedia*, "The cocaine evil was met by stringent legislation in Massachusetts, Ohio, Oklahoma and New York."

In New York City, Seventh Avenue was the home of the cocaine trade. In May, shortly after the enactment of the city's Cocaine Act, one Doc Lewis was arrested and convicted for selling cocaine. It was his fourth arrest and second conviction for drug dealing. He was sentenced to two years in state prison, one of the severest sentences yet handed out under the new act.

The next day in Oklahoma City, a "coke fiend" shot the sheriff. Apparently the fiend had not taken advantage of any of Oklahoma City's drug abuse sanitariums, like White's,

• • • • • • • • • • • • • • • •

which advertised, " a guaranteed cure for Whiskey - Morphine - Opium - or Heroin addiction."

The "Pleasures" of the New York Subway

In 1910 the New York City subway system carried over 200 million passengers, 80 million less than the Paris Metro. The New York system was already known for its congestion. In a *New York Times* article called "The Subway Crush Causes Mental Strain," Dr. Carlos MacDonald, an expert on nervous disorders, complained, "I have found myself literally fighting my way through the crowd in the cars for several minutes before the station is reached, and repeating the struggle in making my way through the crowd on the platform." MacDonald concluded that the strain of riding the subway was a leading cause of mental disorders. "Many foreigners on first visiting New York," he continued, "remark the peculiar and abnormal set expression on the faces. It is pointed out again that a large proportion of New Yorkers talk to themselves on the streets, and that this is an evidence of a highly nervous condition. . . . As the subway crush becomes a habit, its injurious influence become less serious. One can become accustomed to almost anything."

Crime in New York City

The New York crime rate was unusually high, and a frequent target of complaints and investigations. Criminal gangs fought pitched battles with police on the streets of the city. The police department itself was full of corruption, which would break into a public scandal in a couple of years.

In mid-May, goaded by the incessant public outcry over lax police action, Mayor Gaynor addressed the inspectors of the New York Police Department: "The only thing I want is that you should go out and do good, plain, ordinary police work. In other words, I should like to have each citizen in New York feel that he can go to bed with a fair assurance that his throat won't be cut before morning."

. . . .

Before the year was over, Mayor Gaynor was the target of an unsuccessful assassination attempt.

A Woman's Place in 1910

The battle for women's suffrage was the major political issue for women in America and England. Norwegian and other Scandinavian women already had the right to vote, but in the English-speaking world the vote was still withheld.

In the struggle to win equal participation in the political process, other issues were raised that are still with us. One of these issues was clearly stated in England—women were paid less than men, often for the same or similar work. In addition, they were restricted in their access to training, thus limiting the possibilities of advancement.

Mother's Day, denounced by some as a sentimental sop thrown to women to make them forget their second-class citizenship, was fast becoming a national custom. But more to the point, widely read columnist Dorothy Dix was advocating salaries for wives: "He (the husband) works eight hours, she works eighteen, but few wives would be able to get more than an IOU."

On the educational front, women were calling for more access, but in some places the tide seemed to be running the wrong way. In May, feminist Julia Ward Howe, called "the greatest champion of her sex in the world," condemned the action of the trustees of Tufts College in Medford, Massachusetts, who converted the school from a coeducational institution to a college for male students only.

At its June commencement, Jane Addams became the first woman to receive an honorary degree from Yale. That year Addams published *Twenty Years at Hull House*, hailed by some reviewers as "the most important book of the year." They had in mind the far-reaching implications of the work Addams had done at her Chicago Settlement House, where she practically invented social work and social welfare as a department of life in the United States.

On May 21, two days after our passage through Halley's tail, the Women's Political Union marched down New York

· · · · · · · · · · · · · · · · ·

City's Fifth Avenue. Only a few hundred demonstrators appeared on this cold, drizzly day, but it was the beginning of a march that would lead to women's full participation in American political life. A few weeks later, the women of England who had already wreaked physical and emotional havoc in Parliament, staged the biggest demonstration yet. Ten thousand "Suffragettes" marched from Victoria Embankment to Albert Hall.

The narrow hobble skirt appeared in 1910 and soon became the look of the day. The tightness of this floor-length narrow sheath limited a woman's movement to steps of only about six inches. It was an ironic oddity that in an age of militant feminism, so many women should adopt such a stifling style. This led one commentator to call the hobble skirt's success a "victory of fashion over reason."

In Paris, Auguste Rodin, the greatest sculptor of the age, unabashedly expressed his admiration for contemporary women. "Ancient women were beautiful," said Rodin, "but women of today are their equals." But an academic from Worcester, Massachusetts, gave voice to a widely held masculine opinion. Dr. Max Baff, of Clark College, revealed to the press that "the women of today are no better, from a psychological standpoint, than the savages of old." He pointed to "a woman's love of adornment" as proof of his theory. Although he was opposed to women's suffrage and was a bachelor, Dr. Baff claimed not to be a "woman hater."

Coffin Nails

Junie McCree, a vaudevillian best known for having composed "Put Your Arms Around Me, Honey," a hit song of 1910, also coined the term "Coffin Nails" for cigarettes. Tobacco was very much on the national mind in that year, with over 8½ billion cigarettes sold in the United States. Tobacco companies spent over $18 million on advertising. Some doctors were suggesting cigarettes to nervous or overweight patients, but other doctors were already convinced that smoking was dangerous.

Even some nonsmokers were aware of the danger, and tried

• • • • • • • • • • • • • • •

to protect themselves. In the middle of May, in New York
City, a group of private citizens organized the Nonsmokers
Protective League and made a futile attempt to curb smoking
in public places. And in Oklahoma City, the sale of cigarettes
was banned.

Handling Radioactive Materials

A concern of the day was the safe transportation of radium.
In 1910 a "safe" container was invented. Radium-barium
chloride was enclosed in a so-called "radium cell" some two
inches in diameter and three-quarters of an inch long. The
cell was enclosed in a tube of brass provided with a lead
bottom. A mica plate on one side of the capsule did away
with the necessity of opening it when in use. As one magazine
optimistically remarked, "With new facilities for the safer
transportation of radium, the trade has been placed on a
firmer footing."

If Halley could see, it would probably be bored with Earth
by now. Some places never seem to change. But one of Earth's
enduring foibles is flattering to Halley, and that is our unending fascination with it. We wait for it, we watch for it, we
create theories to explain its origin and behavior. But most of
all, we carry an image of the comet deep within us, where it
often slumbers, but never disappears.

Chapter 15

Comets, Collisions, and Dread

Comets work on the human imagination in ways we do not understand. They stimulate a deep-seated response by tapping into what seems to be an ancient and subconscious dread. What we *know* about comets does not seem to change how we *feel* about them. It is not surprising that fantasies of destruction caused by comets have played such a large role in science fiction, nor is it surprising that so many writers have attempted to prove that the appearance of a comet always coincides with disaster.

By long-standing tradition, comets have been used by science-fiction writers as agents of destruction. The list of novels, stories, films, and television shows that have used comets in this way is long indeed, and is likely to continue growing. Dragons and other monsters provide another source for science-fiction adventures, and our unending interest in them has been traced by some observers to hidden memories of real-life creatures. This theory has also been applied to our fear of comets.

Immanuel Velikovsky, the controversial twentieth century scholar, has suggested that the entire human race carries a subconscious memory of cataclysms caused by actual en-

(From the Zwerdling–Miranda Collection.)

counters with comets thousands of years ago. If such a subconscious memory actually exists, it would provide an explanation for the outbreaks of comet fever that have occurred throughout history. The most recent of these took place in late 1973 and early 1974. The immediate cause was the announcement of a new comet that would visit our skies, comet Kohoutek. Before the comet even came into view, numerous comet-cults sprang up. On December 21, winter's darkest day, a cult known as The Children of God published a cautionary pamphlet that announced the end of the world on January 31, 1974. "40 DAYS! and NINEVEH SHALL BE DESTROYED!" blared the title, and beneath it was a diagram of Kohoutek's orbit labelled with the catastrophic question. "WHAT TERRIBLE EVENTS WILL THE COMET BRING?"

Fanatics were not the only people whose imaginations were gripped by this comet. A milder version of the fever struck the general public. During a three-week period, the American Museum of Natural History in New York was deluged by more than 42,000 telephone and mail inquiries about comet Kohoutek. This does not include the thousands of inquiries addressed to police, universities, newspapers, and radio and television stations. And this was only the New York City response. One of the most interesting aspects of Kohoutek fever was the fact that the comet itself was a dud: it was dim, hardly visible, and generally disappointing. The public response, as it turned out, was to the *idea* of the comet rather than to the comet itself. This episode lends support to Velikovsky's theory of a universal and subconscious comet phobia.

The Kohoutek episode also provides a forewarning of what we should expect during Halley's current visit. Speaking of Kohoutek, the usually reserved *Scientific American* said, "There is no reason to believe the public response in 1985–86 will be less enthusiastic for Halley, the most famous and best known of all comets."

Velikovsky's ideas have been widely ridiculed by the scientific community, and he has been denounced as a writer of science fiction rather than of science. Nevertheless, his ideas and theories do not seem to die. Contemporary scientific investigation has tended to support some of Velikovsky's ideas

. . . .

about the role comets have played in the evolution of Earth and earthly life, though most scientists would prefer not to be linked with him in any way.

The Nemesis Theory

In the summer of 1984, the International Astronomical Union sponsored a symposium at Boston University on "The Search for Extraterrestrial Intelligence." One of the topics discussed was the possible effect on earthly life of an encounter with a comet. The comet scenario developed at this conference is composed of several elements. The first was provided by Nobel Prize–winning physicist Luis Alvarez and his son Walter, a Columbia University geologist. They discovered extremely thin layers of iridium-rich clay that had been deposited millions of years ago in ancient strata of sedimentary rock. Geological investigations indicate that the iridium layers were deposited at intervals of roughly 26 million years. The Alvarezes believe the iridium is not of earthly origin. What's more, they believe it came from comets which struck the Earth.

Following the implications of the Alvarez's work, other scientists investigated impact craters caused by comets in the distant past. This study revealed a similar pattern: comet impact craters seem to have been created on the same 26-million-year cycle.

Now the search turned to the significance of the time cycle —what was the cause of the 26-million-year period? Astrophysicist Richard Muller of the University of California and the Lawrence Laboratory at Berkeley put forth a startling hypothesis, the "Nemesis Theory." Nemesis * is the name of an

*As this book goes to press, Dr. Muller and his colleagues are conducting an intense search for Nemesis. They believe they will find it, and argue that it has not been found before only for the reason that nobody has looked for it. Their search was spurred in the fall of 1984 by a major discovery: the existence of a planet outside of our solar system, a planet which orbits around another star (the planet VB 8B, which orbits the star known as Van Biesbroeck-8). This discovery demonstrated two things: the existence of other solar systems, and how much we still have to learn.

as yet undiscovered star which, according to the theory, travels through space as a dark companion of our own sun. Nemesis and the sun are believed to orbit around each other in a relationship of two stars known as a binary-star system. Nemesis is thought to be a "red star" of about the 10th magnitude (which means it is 100 times too dim to be seen with the naked eye). Furthermore, Muller theorizes that this dim, obscure star travels an orbit which brings it relatively close to the sun every 26 million years. When it is nearby, Nemesis pulls comets from the outer fringes and into the heart of the solar system, where Earth resides. The result is an increased probability that Earth will be hit by a comet.

The close proximity period of Nemesis is thought to last for about 2 million years, during which time a dozen or more comets might strike Earth. "We must conclude that you're going to have periodic impacts," said Professor Muller, "and the comets must be coming in showers."

According to this scenario, comet collisions have had a great effect on earthly life, with particular emphasis on the process of evolution. The impact of a single comet can be enough to disrupt Earth's biosphere for years and wipe out entire species of animals. Many contemporary scientists believe a comet wiped out the dinosaurs. One of these theorists, Dr. J. Sepkoski of the University of Chicago, believes comet collisions or some other catastrophe are required for evolution to occur: "Extinction appears necessary to make room for new species to evolve." *

*The role of comets in the evolutionary process was dramatically brought to the public's attention in the spring of 1985 when the cover of *Time* magazine (May 6) was devoted to the story. The *way* in which the story was told struck a familiar chord: it featured much of the mixed-message hype that often surrounds comet stories. Tyrannosaurus rex flees across the cover in fear. Overhead, a half dozen comets streak toward Earth. On the far horizon we see what has terrified Rex: an ominous mushroom cloud rising out of a flaming explosion. Across the cloud, in bold red letters, the title of the cover story: "DID COMETS KILL THE DINOSAURS? A New Theory About Mass Extinctions."

The article inside focuses on the 26-million-year cycle of comet destruction
(footnote continues on page 150)

. . . .

When a comet strikes Earth, the impact may gouge out a crater as large as eighty-five to ninety miles in diameter (two craters of this size have been found, one in Canada, the other in South Africa). In addition to the destruction caused by the initial shock wave and the associated lethal heat, poisonous gases, tidal waves, and earthquakes, there is a long-term effect. The material that formerly filled the crater is hurled into space, slowly drifts around the world, and falls for years as fine dust. (This is when the iridium layer is laid down.) During the dust fallout period, the sun is obscured and the Earth is enclosed in darkness that might last for months or years. Without sunlight, photosynthesis, the primary process in the food chain that supports all life, comes to a stop. Some species starve to death. Others, with smaller and more adaptable feeding habits, might manage to find enough to eat, but the loss of sunlight also leads to the loss of heat, and many of them die in the ice age that follows. (The aftermath of a comet collision sounds very much like a nuclear winter.)

The dinosaurs disappeared 65 million years ago. If a comet collision caused their extinction, and if the collision cycle is 26 million years long, there were two comet showers since then; the first 39 million years ago, and the second 13 million years ago. Scientists have found some evidence to document these two most recent showers. What it means for Earth is unavoidable: another comet collision of major proportions is in our future. But there is no need to start worrying yet, we still have about 13 million years to go. And while we fear future collisions, we have reason to be thankful for those of the past. As Professor Stephen Jay Gould of Harvard remarked, "If it hadn't been for the cometary showers, or whatever, that removed the incumbents, we wouldn't be here today."

---

and offers the Nemesis theory as one explanation. We are reassuringly told that we are now in the middle of a cycle, with 13 million years to go, but, *Time* adds, "It is also possible, *even probable*, that long before that time, astronomers will spot a random, incoming comet on . . . *a direct collision course with the earth.*" (Emphasis added.)

When Did a Comet Last Hit Earth?

Several scientists say we experienced a small-scale cometary impact earlier in this century. In the early-morning hours of June 30, 1908, the crew and passengers of the Trans-Siberian Express witnessed an awesome sight. A huge ball of intense blue flame streaked across the morning sky, trailing smoke behind it. A few moments later, above the clatter of the train, they heard a thundering explosion.

Reports of this event intrigued scientists around the world, but because of travel restrictions imposed as a result of the Russian Revolution and World War I, investigators did not reach the site until almost twenty years later. When they did, they found that the explosion had occurred at Tunguska, in east-central Siberia, 600 miles distance from the train when the crew and passengers heard it. Thousands of square kilometers of Siberian forest had been flattened by the shock wave. All the fallen trees lay pointing away from the epicenter, like millions of spokes on a giant wheel. The skeletal remains of thousands of reindeer lay among the debris. But what they didn't find was even more interesting . . . there was no crater.

If the blue fireball had been a meteor, which are composed of dense, heavy matter, it undoubtedly would have created a large crater. The missing crater was mystifying until it was linked to Dr. Fred Whipple's "dirty snowball" theory of comets. This theory says that the head (and nucleus) of a comet is composed of ice and small particles of matter. If a comet is indeed mostly ice, the mystery of the missing crater of Tunguska could be solved: the comet or piece of a comet* was relatively small and melted completely as it streaked through the Earth's atmosphere. The friction of this passage would have produced intense heat. What hit the Earth was a speed-

---

*It has indeed been theorized that a very large comet exploded sometime before 1178 A.D. According to this theory, the comet which we now know as comet Encke, which has the shortest period of all known comets (3.3 years), and the Tunguska comet are fragments of the larger comet which exploded.

. . . .

ing mass of steam, superheated gases, and small particles of solid matter.

In support of this interpretation, an analysis of the known data about the Tunguska event has yielded the following numbers: the Tunguska comet was about 120 feet in diameter and weighed in the vicinity of 50,000 tons. The major explosion occurred about thirteen or fourteen miles above the Siberian forest, with a force of 12.5 megatons.

A comet with a diameter of 120 feet is relatively small, which would account for the fact that it completely vaporized before it hit the earth. However, the average comet that strikes the earth has a diameter of about half a mile and hits with a force of 100,000 megatons of TNT. A megaton is a million tons, which makes such an explosion equal to the force of one hundred thousand million tons of TNT. It has been estimated that the probability of a collision with a comet of this size is once in every 260,000 years.

Probability also points to a collision with a six-mile-diameter comet approximately every 100 million years. The force of an impact of this magnitude would be equal to 1-trillion tons of TNT.

To place the destructive power of such comet impacts into perspective, by 1985 the cumulative nuclear arsenal of the entire world amounted to no more than 10,000 megatons.

Nuclear power is one of the new things Halley will witness. Another new addition is perhaps even more startling. This time around, Halley won't just be visiting Earth, we will visit it.

Chapter 16

Destination Halley

In the spring of 1986, for the first time in history, space probes will rendezvous with a comet. There are several compelling scientific reasons for Halley to be the first comet to receive this high-tech treatment—its size and its known history, for example. But there is also another compelling reason, one based on human emotions. As the most famous of all comets, it is only fitting that Halley should be the object of a series of space studies which will, in all likelihood, provide more hard data about the nature of comets than all the studies that have been conducted in the past.

Five space vehicles, launched from different parts of Earth, have been sent to meet Halley. Three of the probes are specifically designed to study and photograph the comet's nucleus and one of them will approach to within 350 miles. The other two probes will operate at a greater distance. They will observe the comet from about 60,000 miles away, roughly just beyond the limits of the hydrogen envelope that enshrouds the comet's head.

The space probes shall be sent by the Soviet Union, Japan, and the European Space Agency (ESA), a consortium of eleven nations: Belgium, Denmark, France, Germany, Ireland,

A model of Giotto, the probe sent by the European Space Agency to rendezvous with Halley in 1986. (Photographed by Gabrielle Rossmer.)

. . . .

Italy, the Netherlands, Spain, Sweden, Switzerland, and the United Kingdom. The United States is most noticeable by its absence from this list. American scientists had originally planned to send an elaborate probe of their own. But the funds were not forthcoming. According to one scientist at the Jet Propulsion Lab, who preferred to remain anonymous, "The decision not to fund an American space probe was unspeakable and terribly insane."

Vega I and *II*

On December 15, 1984, at the Baikonur Space Center in the steppes of Kazakhstan, the Soviet Union launched the first spacecraft ever to be aimed at a comet. *Vega I*, weighing 8,800 pounds and carrying among its equipment two television cameras and an assortment of scientific sensors, was blasted into space atop a huge Proton rocket. Thus began the most exciting scientific research of Halley's current visit. A short time later the sister ship *Vega II*, was also successfully launched.

The name "Vega" is a condensed version of the Russian words for Venus (*Venera*) and Halley (*Gallei*). Venus is part of the name because it is the first destination on the route of these probes. When they reach Venus in June of 1985, they are scheduled to deploy a landing craft and an atmospheric test balloon. Then they will take advantage of the planet's gravitational field by using it as a "springboard" to gain energy, change course, and continue on towards Halley.

Vega I is programmed to fly through the dust cloud, or coma, around the comet's nucleus, and will come to within 6000 miles of the nucleus during the first week of March 1986. On its flyby, it will attempt to obtain the first photographic images of the nucleus. It will also obtain and relay information to Earth on the dust cloud surrounding the comet. Based on the information gathered by *Vega I*, flight controllers on Earth will determine the closest possible flyby distance for *Vega II* which, under ideal conditions, will pass to within 1800 miles of the nucleus.

. . . .

Despite being known by their countries of origin, the various probes represent a significant international, collaborative scientific effort. When NASA was denied funds to mount its own probe, the Soviets promptly contacted Dr. John Simpson of the University of Chicago. In the fall of 1983, Dr. Simpson had delivered a paper in Holland on a new technique for measuring cosmic dust. As a result of the paper, the Soviets requested his assistance on their project.

With no time to spare, Simpson and his colleagues received the approvals they needed from the White House, the State Department, the Defense Department, and NASA. Once they had the green light, the technical problems had to be addressed. All of the equipment had to have compatible electrical and computer connections. In effect, the equipment had to speak the same language. As of this writing, the equipment has been launched. It is working perfectly, and data is being continuously sent from the satellite to Soviet receiving stations on earth; simultaneously, Simpson's data is being fed into a special TELEX hotline, where he receives it in Chicago for analysis.

Simpson's dust analyzer is not the only American contribution to the Vega effort. Americans worked on the plasma physics experiments and on at least one of the craft's spectrometers. Bradford Smith of the University of Arizona contributed to the imaging equipment on board the spacecraft. Early 1986 will be a momentous time in his career. In January, two months before *Vega* flies by Halley, NASA's *Voyager 2* will fly by and send back images of the planet Uranus. *Voyager* will be using Smith's imaging equipment. Ironically, his equipment could possibly discover that Uranus is actually a very large comet, one of the theories that has been proposed by astronomers.

Giotto

In June of 1985, astride a European-built Ariane rocket, the ESA's space probe known as *Giotto* was launched from a rocket base in Korou, French Guiana.

. . . .

In choosing the name *Giotto*, the ESA paid homage to Giotto di Bondone, the fourteenth-century Italian painter who depicted the Star of Bethlehem as Halley's comet in one of his masterworks, "The Adoration of the Magi." The craft is scheduled to make its Halley's flyby in mid-March 1986, a few days after *Vega II's* passage.

Giotto is capable of passing within a remarkable 350 miles of Halley's nucleus. If all works properly, it will send photos and data back to Earth over a brief four-hour period. But by the time *Giotto* has been on for four hours, it will probably have been pummeled and sand-blasted to bits by the onslaught of cometary dust.

What a four hours they will be! *Giotto* is packed to the limit with some of the most sophisticated equipment ever sent into space. Among these technological wonders, as listed by the ESA, are:

- A camera for imaging the inner coma and the nucleus;
- Neutral, ion and dust-mass spectrometers for composition measurements;
- A dust-impact detector for studies of the dust environment;
- Various ion plasma analyser, an electron analyser, an energetic-particle experiment and a magnetometer for studies of plasma physical processes;
- An optical probe for in-situ measurements of the cometary dust and gas environment.

What will we learn from *Giotto*? If all goes well, scientists will obtain data to help them solve some fundamental cometary mysteries. One is the chemical composition of the coma and cometary dust; and another is to determine the dust-to-gas ratio within the comet's head. In addition, *Giotto's* cameras will be able to take pictures of the comet's nucleus with a resolution down to 150 feet. These pictures will not only be beautiful and exciting, but will also enable scientists to determine the size, rotation, and perhaps even the mass of the nucleus itself.

• • • • • • • • • • • • • • • •

Planet A and MS-T5

Two probes, named *Planet A* and *MS-T5*, represent Japan's entry into interplanetary travel. They were launched in August of 1985 from the Kagoshima Space Center atop Japanese-built rockets. Unlike the others, the Japanese probes are not capable of extensive midcourse corrections. However, ground manipulation is less critical since the Japanese probes are routed to station themselves no closer than 60,000 miles away from the nucleus.

From their panoramic vantage point, they will study the head and tail on a wide-scale basis. Among their objectives, they will study the manner in which the head and tail grow and shrink, and perhaps more importantly, they will focus their instruments on the interaction between the comet and the solar wind.

They will watch the cutting edge of the comet as it carves its way through space. This cutting edge has become a fascinating field of study as our understanding of the solar system has increased. The solar wind is responsible for the formation of a comet's tail. But currently no one is certain of the relationship between the solar wind and a comet's coma or the solar wind and Earth itself. *

---

* To help to understand the solar wind, a space project named AMPTE—the Active Magnospheric Particle Tracer Explorer project—released a cloud of barium vapor beyond the Earth's magnetic field early on the morning of December 27, 1984. The press called it an "artificial comet" and presented it to the public as something new. The objective of the experiment was, in effect, to calibrate the solar wind so that a stable measure would exist against which to measure the effects upon comets.

Interestingly, astrophysicists and other observers saw this kind of media coverage as one of the preludes of comet fever. Barium releases into the atmosphere, in fact, have been taking place for over twenty years. By calling the experiment an "artificial comet" and by linking it to the return of Halley, the experiment received extensive national coverage. "Had someone released an 'artificial comet' in 1960," one astrophysicist muttered, "it wouldn't have made the back page of any newspaper."

. . . .

The World Unites Around Halley

The degree of cooperation among twenty or more nations around the comet probe programs is perhaps unprecedented. Every single one of the probes is tied into the other probes on several levels. Scientists from all over the world are participating. In some sense, this visit from Halley is fulfilling the age-old scenario which depicts the nations of Earth unifying against an invader from space.

The success of each of these missions depends, to some degree, on the successful data gathering of the others. Each in its turn will transmit data that will allow scientists to finetune the next shot. The International Halley's Watch plans to have published *all* of the research of over 800 professional astronomers and countless thousands of amateurs by 1989. Surely scientists will still be untangling that massive web of data in 2061.

Professor Emeritus Fred Whipple, *former director of the Harvard-Smithsonian Astrophysical Observatory and a world-class cometologist.* (Photographed by Gabrielle Rossmer.)

Chapter 17

Fred Whipple: The World's Best Comet Watcher

Six different periodic comets and one major observatory carry the name "Whipple." But for the quirks of government bureaucracy, the meteor shields that protect all space capsules, including the ones scheduled to rendezvous with Halley, would be called "Whipple Meteor Bumpers." In addition, and perhaps most importantly of all, in 1950 Fred Whipple created the Dirty Snowball Theory of comets. Most scientists believe this theory will be substantiated during Halley's current visit.

Whipple's accomplishments are as impressive as those of any living astronomer. Some observers have even called him the most famous astronomer since Halley. Along with James Van Allen (discoverer of the Van Allen Radiation Belt), Whipple is one of the two scientists ever to be inducted into the International Space Hall of Fame. For eighteen years he was director of the Harvard Smithsonian Astrophysical Observatory in Cambridge, Massachusetts. On an afternoon in the winter of 1984, he sat back in his chair in his office at the observatory, surrounded by such tools of his trade as detailed globes of the planets and a pair of computers, and he talked about Halley's comet.

. . . .

"I am excited about seeing Halley's comet. I've been waiting a long time. I'm not excited about looking up and seeing a comet with a tail; I am excited about getting good photographs of the nucleus and finding out what compounds are near it. I don't know what a nucleus looks like, but I'd sure like to see one.

"I think there is a good chance we'll be able to see the nucleus clearly. We have a week with three different missions going by and all of them can take pictures. Surely one of them will get something.

"If the space probes work as well as they could, we will learn an enormous amount about comets. We will see what the structure is; we'll see the irregularities, the variations across the surface, the nature of the different materials and different gases coming out of it. We'll be able to see the dust coming out of different zones.

"Right now we have no idea what molecules are at the nucleus. Since the atmosphere acts as a little chemical laboratory, by the time the gas goes out far enough to make a signal that we can receive, it has already been chemically changed. The original materials remain a mystery. I am almost certain we will find new chemical compounds on this visit.

"If the cameras actually do get a shot of the nucleus, it will be the first image we've ever had of one. There is no question at this point that comets have a nucleus. We have observed three of them by radar. Unfortunately, radar cannot help us determine reflectivity or rotation periods so we cannot determine exact dimensions.

"I think the nucleus will probably be black as slate, like a slate blackboard, but not quite as dark as black velvet. The ultimate dimension of the nucleus, in fact, will depend upon how dark it is. If it is very dark, it will be larger than we thought because dark objects absorb so much light and look smaller to the viewer, particularly over a great distance. Our best estimate right now is that it is about four miles. It may be larger than that, but we doubt that it is any smaller.

"I also think the shape of the nucleus will be irregular, with large cracks and valleys, and mesa-type areas. I think this, in part, because of the irregularity of dust output, and partly because I am sure that the nucleus is made up of the influx from smaller comets. Therefore there will be somewhat different compositions and different ices."

During the 1980s, Dr. Whipple has been deeply involved in the planning of some of the space probes. A small, detailed model of *Giotto*, the European comet probe, sat on his desk. Handling it delicately, he pointed to an instrument protruding from one side and continued speaking.

"The part of *Giotto* that I am most interested in is a little gadget on the side that will help us to photograph the nucleus. It is a scanning mirror. As *Giotto* rotates—once every four seconds—this device will assure that the comet remains in the camera's field of view."

Were Comets Responsible for Earth's Atmosphere?

Dr. Whipple does not take the study of comets lightly. He has theorized, in fact, that comets could be responsible for Earth's current atmosphere, the atmosphere in which life developed.

"It is quite clear, and well accepted among scientists, that Earth lost its original, primitive atmosphere. The original atmosphere was probably lost as a result of a high influx of foreign material, probably caused by the very strong, high-energy solar wind. The atmosphere we have today is not a remnant of the primitive atmosphere at all. We base this knowledge primarily on the abundance of the noble gases: krypton, argon, and so forth.

"We've known that much for a long time. What we don't know is how Earth got its present atmosphere. The geologists have a theory and I have a theory. The geologists believe—and they are probably right—that the gases seeped out very slowly from inside the Earth.

"But there is also the off-chance that the atmosphere came from comets! No one has disproved it, so it might be true. I'm certain that comets landed during the period after Earth lost its primitive atmosphere. What they brought with them or whether or not the amount was significant, nobody knows. *I*

• • • • • • • • • • • • • • •

*am sure, though, that some of the atoms in your body came
from comets."* *

Dr. Whipple Makes a Theory . . .

As we have repeatedly seen, the amount of hard knowledge
we have obtained about comets remains relatively scant. We
observe them; we compute their orbits; we spectroscopically
analyze their light; and we apply whatever other tests are
known to science. Nevertheless, the unusual behavior of comets poses problems that have defied solutions.

Like Dr. Whipple and his astronomical colleagues, scientists who work on the cutting edge of any field of research are
frequently put in the position of having to develop theoretical
models. These models are created to help us understand details of a phenomenon which we do not yet know. Hence, the
Dirty Snowball Theory of comets.

"In the 1940s it had become clear to me that the Swarm
Theory could not be true; there was entirely too much gas
coming out of comets to permit them to survive for a great
many revolutions around the sun. But I had shown that

---

*Astronomer Sir Fred Hoyle has theorized that all life began in comets and
that diseases still come from them. According to this theory, comets received
the carbon-based chemicals necessary for life at the moment the universe
began with the Big Bang. Thereafter, a long history of comet collisions
sparked various chemical reactions. At first, Hoyle thought the comet material melted as it passed by the sun and then remained on its passages through
space in a kind of womb inside the comet. More recently, Hoyle has proposed
that radioactive aluminum in the comet could generate the required heat
internally. Regardless of its source, according to this theory the action of heat
on the chemical combinations inside comets created the basic forms of life.
Over long periods of time, the deep freeze of outer space refroze the newly
formed bits of life, and there they drifted until they again thawed, this time
on the surface of the Earth or some other planet.

This theory has not been well received by the scientific community. In a
professional review of Hoyle's ideas, Dr. Whipple wrote, "As for Sir Fred's
proposal that life originated in comets, I am charmed but not impressed by
the picture of life-forms developing in 'warm little ponds' protected in their
icy igloos from the cruel cold and near vacuum of outer space, falling to Earth
at speeds exceeding 25,000 miles per hour."

Encke's comet had made hundreds, perhaps thousands, of revolutions. There was no way it could be a "flying sand bank" as the theory of the time said it was. There *had* to be a reservoir of gas.

"That idea got me going, but the thing that really sold me was that Encke has a nongravitational motion. It defies the law of gravity. It was coming back about two and a half hours earlier on each revolution. It's period was getting shorter. That can be accounted for by the jet action, or thrust, of evaporating gases!"

What this meant to Whipple was that a large volume of gas was escaping from the comet, and if, in fact, the comet were a flying sandbank, the gas would completely escape in only a few revolutions. Therefore, putting his facts together, he had to create a model which would provide a reservoir of gases, only a small quantity of which would be released on each orbit. The result was the Icy Conglomerate Model.

"It was probably somebody in the press who coined the term 'Dirty Snowball.' I don't know if I ever used it myself or not. But it is a very literal and accurate statement. Comets are 'dirty snowballs' made up of a huge amount of extremely fine dust, mostly water ice, and some other very sophisticated ices."

Dr. Whipple no longer performs astronomical observations; he leaves the telescopes to his younger astronomical colleagues. As for seeing Halley, he advises the public to use binoculars, not a telescope. "Two eyes," he says, "are better than one."

Most of Whipple's effort on this current visit will go into various aspects of the space probes, but he will surely get the best look at Halley he possibly can. He was four years old during its last visit, but he did not get to see it. His attitude toward comets is very personal.

"There really is no telling what any specific comet is going to do. They are all individuals, just like us. They are even sometimes entirely different on each visit. I don't think the comet will be much brighter than the Andromeda Nebula, but I do think there will be a good tail."

. . . .

Dr. Whipple smiled, as if pleased that the comet might put on a good show for Earth. "But about that comet fever stuff," he said, "there's a lot of hype already, and there will be a lot more. There's no telling how big they can make it."

The "they," of course, are the media. The hype got off to a quicker and bigger start than it did for the 1910 visit. Beginning with Halley's recapture in late 1982, the comet appeared in print with increasing regularity. By early 1984, it was showing up everywhere, from small news items off the wire to feature coverage in the *New Yorker*, to the syndicated column of William Safire, who admired the comet's dependability: "Halley's Comet: Something to Be Counted on in an Uncertain Age."

The electronic media, with their longer leadtime, got started even earlier. In 1984, one of the former astronauts taped a series of TV and radio spots and they were being marketed to various outlets. By November of 1984 the comet market had grown so bullish that it was featured on the front page of the *Wall Street Journal*: "Although the Comet Is Still a Year Away, Halley Hype Is Here." The headline referred to the rapidly expanding inventory of Halley-related merchandise. As the subhead succinctly put it, "Promoters Push Telescopes, T-shirts, Pills, Medallions and Cruises to the Tropics." Other comet items for sale in 1984, a year before the comet would arrive, included sweatshirts, windbreakers, rollbags, hats, caps, bumper stickers, reproductions of 1910 comet postcards, comet stock certificates, wall charts, posters, drinking glasses, newsletters, balloons, cardboard models of the comet's orbit, pins, medals, and packaged tours to locations on Earth which were thought to offer the best views of the comet. All of which leads to the question: If hype comes, can predictions of disaster be far behind?

Chapter 18

Who Was Nostradamus and What Was He Telling Us?

As they have always done, the world's astrologers will factor Halley into their predictions. Whatever meaning contemporary astrologers attach to the comet, they will undoubtedly revive the words of Nostradamus.

Nostradamus. The very name itself is wrapped in mystery. An oracle, a seer, perhaps the most famous name in all astrology. Nostradamus, an eerie voice from the past. A voice with something to say about almost everything, including Halley's comet in 1985–1986.

Nostradamus, born Michel de Nostredame in France in the year 1503, was a physician and an astrologer. He first came to fame during an outbreak of the plague in southern France. It was said he worked miraculous cures.

Soon after, he was called in by the Inquisitors under suspicion of heresy. He chose instead to flee, and wandered through the countryside for several years. This was when he began to write down his prophecies, and soon published them in small almanacs.

His almanacs found favor with the reading public, and he set himself the task of writing a series of prophecies which would encompass all events between his day and the end of

(From The Library of Congress.)

· · · · · · · · · · · · · · · ·

the world, which he believed would occur in the year 3797. The result of his labor was *The Centuries*, a poetic prophecy written in quatrains. The title of the work stemmed from the arrangement of the quatrains into sets of one hundred.

In Century II, two quatrains, numbers 43 and 62, refer to Halley's comet. In a recent edition of the works of Nostradamus, the editor notes that Halley's 1985–86 visit will once again have a profound impact on Earth and its inhabitants.

This is what Nostradamus predicts. The translation is literal:

Quatrain 43

In the Time when the Hairy Star is in View,
The Three Great Princes shall become Enemies,
Heaven will send Down Earthquakes,
Tidal Waves shall heave Serpents onto the Shore
    from Arne and Tiber.

Quatrain 62

Mabus shall arrive, and soon after shall Die,
There shall be a Horrible Destruction of People and
    Animals,
Suddenly the Vengeance shall be Seen,
Blood, Thirst, Famine when The Comet Flies By.

One commentary on these prophecies says, "The coming of the comet shall occur in the period of reconstruction, and there will be vengeance for wrongs inflicted on humanity by selfish interests."

But what is Nostradamus telling us? How do we interpret his cryptic warnings? His remarks raise more questions than they answer.

In Quatrain 43, who are the Three Great Princes? Is this an allusion to the Big Three world powers? If so, who are the Three? Can we assume that the United States and the Soviet Union are two? Is China the third? or Japan? Once the guessing begins, there are many alternatives.

. . . .

Whoever the Three Great Princes are, what will happen when they become enemies? Does this mean war, or will the three merely isolate themselves in stony silence?

Are Arne and Tiber the Arno and Tiber Rivers in Italy? Does Nostradamus mean to be specific? Is he telling us that comet-caused earthquakes, and the attendant tidal waves, will destroy half of Italy? Or are Arne and Tiber metaphors for all the waterways and oceans of the world? He's pretty clear about the earthquakes and tidal waves, but will the destruction be localized and specific, or world-wide and general?

Quatrain 62 poses its own questions. Who is Mabus? What is the vengeance that shall be seen, and who will send it? God? Halley's comet itself?

Astrology is a method of divining the future based on the theory that the stars and other celestial bodies influence our lives and determine the outcome of events. The precise origins of astrology are lost in the shadows of history, but we know it is very old. Sophisticated astrological systems were used by the ancient Egyptians, Chinese, Babylonians, Assyrians, and other early civilizations. The collection of celestial observations, originally required for astrological purposes, eventually gave rise to the science of astronomy.

In our modern, enlightened age, astrology is deemed a "pseudoscience." Its system and predictions are not verifiable by rational proofs. But astrology has numerically more adherents today than it has ever had before. In addition to the books and periodicals devoted to astrology, columns of astrological advice appear in numerous respectable newspapers. We can be sure that the warnings of Nostradamus, whatever they mean, will be widely circulated during Halley's 1985–1986 appearance. And we can also be certain that many events will be interpreted to fit his predictions.

Conclusion

On June 16, 1911, more than a year after Earth's encounter with its cyanogen tail, astronomers caught their last glimpse of Halley as it disappeared beyond telescopic range into the vast distances of the Solar System. Halley came, Halley went, and Earth was still here.

Comet-related suicides and accidents had claimed some lives; here and there some poor souls had slipped into insanity; but the overwhelming majority of Earth's inhabitants had survived. Many observers smugly proclaimed, "Nothing happened." They were wrong. Something had happened. Virtually every person on Earth, one way or another, had acknowledged Halley's comet. On the night of the comet's tail, the entire human race had participated in a shared experience on a scale unprecedented in history.

On its 1910 visit, Halley soared by an Earth that seemed aimlessly spinning between the end of one era and the start of the next. The world order brought into being by the great powers of the nineteenth century was ready to collapse. Our own era, and the twentieth century itself, as we have come to know it, would begin four years later with the start of World War I.

(Photographed by Donald Gropman.)

• • • • • • • • • • • • • • • • •

As we have seen, many things have changed and many things have not since Halley last was here. What we don't know is if there has been any change in our attitude about the comet itself.

After Halley disappeared in 1911, it was out of sight for seventy-one years, until Cal Tech astronomers photographed it in 1982. It has been in our consciousness since then. The "Halley hype" was well underway by 1983. Newspaper and magazine articles began appearing with more frequency. Comet memorabilia hit the marketplace. Comet tours were advertised. In many ways it began to sound like 1910 all over again.

There is no telling if comet fever will erupt in 1985–1986. We think of ourselves as modern people, the most sophisticated in history. That's precisely what the people of 1910 thought, but that year Halley brought on a worldwide panic.

In 1910 Halley arrived at the start of the Air Age. This time around we are in the Space Age, and will send spacecraft out to rendezvous with the comet. We can safely say that the advances in technology (space flight, television, computer-enhanced imagery, and a host of other innovations) will help us to learn more about comets during Halley's current apparition than all of the knowledge we have gained in the past.

The last major and definitive *fact* learned about comets was Edmond Halley's discovery that they are orbital. Since then, scientists have frequently theorized about the origin and physical composition of comets, but have come to no definite conclusions. In 1985–1986 we will probably discover the answers to many basic quesions . . . How big is Halley's comet? What is the nature of the nucleus in its head? What is the origin of Halley's comet? Perhaps we will also unlock some of the mysteries of the comet's tail.

Scientific understanding of the comet will be increased manyfold, and the international TV audience may see close-up footage of the comet transmitted back to Earth by the Soviet Union's *Vega I* spacecraft, among others. But a question remains. As our knowledge increases, will our dread decrease? Will definite fact drive out vague fears? We cannot

tell. If we look at the long history of the human race's fascination with comets, we are left with doubt. So far, nothing
we have learned about comets has made us less vulnerable to
their mysterious force.

Index

• • • • • • • • • • • • • • •

. . . .

• • • • • • • • • • • • • • •

• • • • • • • • • • • • • •

• • • • • • • • • • • • •